Findlay's Lighthouse Lists and the Loss of the *Margaret Smith*

By Ian Hills

Copyright ©FWD Publishing 2020
All Rights Reserved
ISBN: 9798655453067

Contents

	Contents	3
	Acknowledgements	4
	Preface	5
	Introduction	7
Chapter 1	The Development of Trade and Shipping	10
Chapter 2	The Sailing Ship	12
Chapter 3	The Trackless Ocean	22
Chapter 4	The Unlit Coast	37
Chapter 5	What Light is That?	52
Chapter 6	The Beginnings of Findlay's Lighthouse Lists	59
Chapter 7	The Coasts are Lit	68
Chapter 8	The March of Technology	77
Chapter 9	Warning by Sound	89
Chapter 10	The Last Voyage of the *Margaret Smith*	97
Chapter 11	Alexander George Findlay	112
	Epilogue	115
	Bibliography	118

Acknowledgements

Many people have assisted in this project which has grown from very small beginnings. The staff at Greenock Library went to great lengths to find material about John Kerr and his ships, which helped provide a picture of nineteenth century Greenock. Profound thanks are also due to John McKee and Jonathan R. A. Hill who undertook the arduous task of reading through the original drafts and offering constructive criticism that has undoubtedly led to greater clarity in many places. Once again thanks must go to Chris Hills for providing the photographs and to Jo Nicol for the drawing of the Hartlepool optical apparatus. The contribution of Jonathan Hills, who found many of the references, sorted out issues with the technology and made so many constructive suggestions, has facilitated a far fuller picture and one which is hopefully far more readable.

Especial thanks go to the staff at the Maritime History Archive in St Johns, Newfoundland who were particularly helpful in finding information about the crew of the *Margaret Smith*. Gratitude is also due to the members of the Stocksfield Computer Club who provided invaluable help with some of the drawings. However, this book would never have been written without the support and encouragement of all my family to whom I am most humbly grateful.

Preface

Born the son of an engineer who spent his life working on canals and docks, an interest in things maritime was almost inevitable. Coupled with a love of history, this grew into a fascination with the development of technology, which initially centred on a particular industrial relic. The River Irwell flows south to Manchester and on its west bank are the long-abandoned remains of what was once Fletcher's Canal. At the head of the canal were the ruins of the old Wet Earth Colliery from which a ditch ran north along the riverbank to what looked like a wooden gate. Many hours were spent exploring the site, climbing down holes and crawling up tunnels, activities which today would give any Health & Safety officer nightmares. More traditional research methods in the library revealed that this was a remarkable piece of work by James Brindley, the canal engineer. Just by using tunnels, a siphon and a waterwheel, he was not only able to drain the Wet Earth Colliery, but use the water flow to move the laden coal barges and even supply the industries of Salford with water, all by harnessing the free power of gravity. A wonder in its day, now abandoned and forgotten; overtaken by the relentless march of technology.

In the twenty-first century our advanced technological society with all its remarkable gadgetry makes it all too easy to forget that it was not achieved at a single step, but by advances each of which built upon what had gone before. In 1999, American engineers moved the Cape Hatteras lighthouse over 950 yards in one piece, an impressive feat by any standards for which they deserve every accolade, but they were not the first to move such a structure and while their project was bigger and they had all the advantages of modern technology, they could also make use of the experience gained by civil engineers as they worked to solve similar related problems. One such engineer was John Murray, who, 150 years earlier in 1841 had moved the lighthouse at Sunderland[1] using very similar methods, thereby laying that foundation of knowledge from which today's engineers can move forward. It is very easy to forget that we owe a great debt to innumerable and oft forgotten pioneers in every field of human endeavour who, by their inventiveness, imagination, farsightedness and hard work, achieved perhaps just one step in the march of technological progress.

Today, if we travel by sea it is usually on a ferry or cruise liner. We expect it to depart at the scheduled hour on the appointed day and arrive with similar precision, delivering us safe and sound. It is unlikely that any thought is given to the likelihood of shipwreck. Expectations were very different in the first half of the nineteenth century. This can be exemplified by our family history which shows that a considerable number of our ancestors chose to abandon life in Britain and seek their fortune in far distant lands. Their letters show very clearly their trepidation about a long sea voyage, trepidation that was fully justified. Two families sailed from Liverpool in 1854 bound for America, a voyage that took over two months and was marked with tragedy. The infant son of one family died on passage, his only memorial, an entry in the ship's papers, 'Buried at Sea.' Shipwrecks were accepted as inevitable in much the same way as we accept that car crashes will happen today, it was just part of life. However, just as continuous efforts are made to make our roads safer by better warning signage, improved road layouts and better vehicle design, so throughout the nineteenth century efforts were made to make sea travel safer. This is the story of one man, Alexander George Findlay and his contribution to that effort and how, despite his best intentions, his labours were sometimes in vain.

[1] Hills, I. (2019). Moving Sunderland's Lighthouse – 1841. FWD Publishing, USA.

Introduction

Alexander George Findlay is not a name which will be familiar to many people. He was a renowned hydrographer whose work made a major contribution to maritime safety. The *Margaret Smith* was a humble sailing ship carrying unexciting cargoes such as coal and sugar, not a ship whose name is likely to strike a chord in anyone's memory. However, the hydrographer and the ship are inextricably linked, for it was the failure to heed the readily available information that Alexander George Findlay supplied that led directly to the wreck of the *Margaret Smith*. This book is about the background to Findlay's work and what it tells us about the development of aids to navigation. It uses the story of the *Margaret Smith* to exemplify why Findlay's work was so important.

In the nineteenth century lighthouses were some of the most technologically advanced structures then built. The incredible efforts to establish towers on wave swept rocks at the Eddystone and Bell Rock grabbed the public's imagination as the immense difficulties faced by those pioneers were relatively easy to appreciate. However, the focus of people's interest was on the building of the structure rather than its purpose. The purpose of a lighthouse was obviously to be a place where a light could be exhibited as a warning to sailors, but when the Eddystone lighthouse was finally built, the light it showed was just the relatively feeble light from some candles.

Lighthouses today are seen by many people as something from a bygone age. They may know that they have something to do with providing a warning for ships, but beyond that there is little understanding. There are several reasons for that lack of understanding and the simplest is that we have little experience of real darkness. Today, we take light for granted; our streets are lit, our homes are illuminated and we can have as much light as we want almost anywhere. It was a very different story two hundred years ago when any light came from a candle flame or the small wick of a lamp. On a night with no moon, you could see almost nothing and even with your lamp, objects more than a short distance away were lost in the inky blackness. It took many years of experimentation to create a lamp that gave out a lot of light and the problem was not fully resolved until the advent of electricity.

Another cause of the lack of understanding is that very few people have any appreciation of how a sailing ship actually worked. Unless you have sailed a yacht or been to sea in a sail propelled vessel, then it is very hard to appreciate that your choice of the direction in which to sail is largely determined by the wind, while if you make a mistake, a vessel driven solely by sail does not have a reverse gear. Yet another issue is that so few people today have been on a vessel that has taken them out of sight of land. The ocean may be called a highway, but unlike a motorway there are no road markings or signposts to guide you. The lack of personal experience of these things makes it hard to understand the absolutely crucial role that lighthouses played in the latter part of the nineteenth century.

There are numerous books that cover every aspect of lighthouse development in great depth. They may be of enormous interest to someone who has a special interest in the subject, but rarely cater for the general reader who simply wants to have a basic understanding. This book sets out to provide that basic understanding by providing a brief but understandable explanation of the issues, the problems and the eventual solutions. Thus, the explanation of how a sailing ship works is reduced to the absolute minimum necessary to make the key points and doubtless those with more experience will be justly able to criticize. The physics behind the lighthouse's optical apparatus is complex and there are volumes which devote whole sections to a detailed mathematical explanation, especially when it relates to maximizing the performance of the system. The explanations in this book are constrained to just the very basic principles.

Findlay's work on lighthouses, though vitally important, was not in any way spectacular. His was the unexciting but absolutely essential back-room work that provided some of the vital tools that the mariner required. The work involved meticulous organization, constant up-dating and a remorseless attention to detail, for a tiny mistake could lead a vessel to disaster. It was a huge workload and a colossal burden of responsibility, but every year from 1861 Findlay prepared and published a list of every lighthouse in the world. The books he created were working documents, intended to be taken and used aboard ships where they were often subject to conditions of wear and damp, an environment not conducive to the survival of any book. Moreover, as the information they contained would be out of date quite quickly, the tendency was to discard last year's copy in favour of this year's edition. It is almost certainly for this reason that very few original copies have survived.

It is the data within the lists which tells the real story. It demonstrates the growth in the number of aids to navigation, the gradual lighting of the coasts around the world and the steady advance of technology. The story has been stopped at the end of the century as the trends in lighthouse development had already been established. This book aims to use the data in Findlay's Lists to illustrate and explain those trends. Although Findlay himself died in 1875, his work was carried on for another twenty-five years, by which time the advent of electricity had begun to have a major impact on the design, construction and operation of navigational aids. Moreover, more rapid communication was changing the way in which information was disseminated. The world as Findlay knew it was changing, if only because the steamship was rapidly displacing sailing vessels as the main cargo carrier.

In order to try to present the information in a comprehensible way, one particular ship has been used as an example. The *Margaret Smith* was just one of many hundreds of similar vessels that carried their cargoes around the world. However, it is her very ordinariness that makes her so special in this context for the issues she experienced would have been those experienced by every other vessel. The trials and tribulations that she suffered in the twenty-two years from her launch in 1857 to her loss in 1879 are representative of the life of hundreds of her sister vessels.

A casual glance at almost any edition of Lloyd's List will show a 'Casualty' section, which all too often told of ships that would never again sail the seas. There was a story of tragedy behind almost every entry. The man who chose to make his career as a mariner knew full well that it carried with it a not inconsiderable risk, for every year hundreds of sailors would end up in a watery grave. It was not only the crews of the ships that were at risk, their passengers and cargoes were equally vulnerable. Everyone appreciated that the losses could be reduced if the coasts were adequately illuminated, but that required advances in technology, the willingness to undertake the very considerable expenditure involved and ensuring that the relevant knowledge was readily available to every mariner. It was this last function where Alexander George Findlay was to make his mark.

What of the *Margaret Smith*? She had, by 1879, been in service for twenty-two years and they had been hard years in which she had made around forty-eight voyages from the United Kingdom to ports around the world and back again. Most of these voyages were to the Caribbean, but she had also sailed to India, Australia and North America. Altogether she had sailed at least 430,000 nautical miles (796,360 kilometres). However, this figure is based on the shortest distance between two ports and as we shall see, sailing ships invariably had to follow a much longer route. There were still profits to be made from a sailing ship in 1879, largely thanks to the fact that outward cargoes of coal were almost always available. It was perhaps ironic, that in many cases the coal they carried would often be used to provide fuel for their competitors, the steamships. An attempt has been made to convey at least a very basic impression of what that last voyage was like for the ship and her crew. After months at sea one can imagine the relief felt by Captain Taylor and his crew when their destination, the island of Mauritius, came into view.

After months at sea and having sailed over 8,000 miles, disaster struck as the *Margaret Smith* was within just a few miles of her final destination. She was not the victim of storm or tempest, fire or foe, but of human frailty. Her story ended in tragedy, the only good news being that no lives were lost. However, her loss was very publicly reported as showing the consequences of failing to make careful note of the information and pay due heed to the advice that Alexander George Findlay's publications offered.

Chapter One
The Development of Trade and Shipping

Goods had been transported by sea for centuries, but as explorers found their way to ever more distant parts of the world, so trade and the volume of goods steadily grew. Colonization of America from the early seventeenth century saw trans-Atlantic trade develop, while in the same period, trade with India was also on the rise. Journeys to either of these distant lands often took months and the ships were quite small with limited cargo capacity. Consequently, the volume of cargo that could be carried was small and needed to be of relatively high value to ensure that the shipowner made a profit.

However, the eighteenth century saw the development of the industrial revolution accelerate, creating a rapid rise in the volume of trade. As the cradle of this industrialization, Britain in particular experienced a huge rise in the demand for shipping. There were manufactured goods to be exported and raw materials, together with food to feed the growing population, to be imported. This in turn provided work for a rapidly growing fleet of merchant vessels. In 1702 the British merchant fleet amounted to approximately 323,000 tons, a figure that had grown to 434,000 by 1760 and 1,986,000 by 1803.[1] Thus, in a period of a hundred years, the size of the merchant fleet had increased around sixfold. Few of the ships exceeded 1,000 tons and 600 tons was around the normal size for an ocean-going ship, although many were much smaller, while the much more numerous vessels in the coastal trade were smaller still. The average size would thus tend to be well under 500 tons so it would not be unreasonable to assume that the 1,986,000 figure represented more than 5,000 vessels. However, in the next fifty years or so the tonnage of sailing ships, most with wooden hulls, would more than double to a figure of 4,204,000 in 1860.[2] It should be noted that the term 'tonnage' is not directly related to either the weight of the ship itself or the weight of the cargo it could safely carry. It was actually a measure of the size of a ship based on certain dimensions and used to calculate the fees that the shipowner would have to pay to use facilities such as harbours. Thus, a ship might easily be able to carry considerably more cargo that her nominal 'tonnage' would indicate. In the days when some lighthouses were privately owned the ship's nominal tonnage would also be used to calculate the fees the shipowner was required to pay to the lighthouse proprietor.

The ships themselves may have developed a little by 1803 but they still relied entirely on the wind for their motive power. They were, therefore, just as vulnerable to the hazards of the sea as they had been a century earlier. It would thus be reasonable to assume that a similar percentage of ships were lost in 1800 as had been lost in 1700, but with the vastly expanded merchant fleet that meant around five times as many shipwrecks with a financial loss that was also at least six times greater. In 1859 losses amounted to £1,500,000, equivalent to about £190,000,000 today.[3] That loss represented a serious problem for it meant increased costs in terms of insurance as well as goods lost, not to mention the general disruption to trade. It also meant a big increase in the number of lives lost. The 'Casualty' section of Lloyd's List invariably noted the loss of vessels and after a storm the list could be extensive. Moreover, this list noted only the known losses, ships which had gone aground and left visible traces or had been seen to sink. However, in severe storms sailing ships could easily be damaged and overwhelmed, sinking beneath the waves without any visible trace, their crews consigned to an unmarked watery grave. Thus, the casualty figures in Lloyd's List are likely to have been a significant under indication of the actual losses. Nevertheless, ships were generally much safer when they were well out at sea, the dangers increased sharply as they approached land. It is some measure of these dangers that between 1876 and 1885 no less than 25,528 lives were lost at sea.[4]

However, ships inevitably had to approach land if only to reach the harbour which was their destination, but the entrance to that harbour was just one small part of the coastline. Unless the captain could find the harbour entrance, then approaching too close to any other part of the coast could easily prove to be fatal. Finding the harbour entrance might be difficult enough during the day unless there were very clear and visible features that the captain could recognize. A distinctively shaped hill or cliff could help, but much of the coastline lacked such obvious natural landmarks. In some instances, an especially tall building such as a church tower might provide a vital recognition point, but neither natural nor man-made landmarks were much use at night.

From the dawn of civilization, man had used the light from a flame to penetrate the darkness, so surely the obvious solution to the mariner's problems would have been to simply place lights all along the coast to mark its position. It sounds simple, so why was it not done? The question might be simple, but the answer is, however, far from simple and requires some understanding not only of the technologies involved in creating a lighting system, but also of the specific properties that such a lighting system required. This in turn requires an understanding of the way in which a sailing ship operated and the problems associated with its safe navigation across the oceans.

---

[1] The Heyday of Sail – The Merchant Sailing Ship 1650-1830 – Conway Maritime Press, London, 1995.
[2] The Ship, The Life and Death of the Merchant Sailing Ship – Basil Greenhill, HMSO, London, 1980.
[3] Lighthouses-The Race to Illuminate the World – Toby Chance and Peter Williams, New Holland, London, 2008.
[4] The Unsinkable Titanic – Allen Gibson, History Press, Stroud, 2017.

Chapter Two
The Sailing Ship

As the eighteenth century drew to a close, the oceans of the world were regularly being crossed by hundreds of sailing ships. They carried goods of all kinds as well as people. In fact, trans-oceanic voyages had been undertaken for almost three hundred years, while voyages across the seas around the coasts of Europe, Africa and Asia had been commonplace for even longer. While there had been developments in the design of ships, the vessel of 1800 retained many of the features of its forerunners in 1500 and was still closely related to ships built over a thousand years earlier. Even the ocean-going ships tended to be quite small as there were few deep-water harbours and cargo handling facilities were primitive. In many cases harbours dried out at low tide and ships had to sit on the bottom, something that required a rigid hull which was very difficult to achieve with wooden construction. Today we expect ships to berth alongside a quay and have their cargo removed by cranes or other mechanical means which can move thousands of tons of cargo in hours. In the first half of the nineteenth century many harbours offered nothing more than an anchorage where cargo was transferred by small boats and everything had to be done manually. As the century progressed more and more docks, where ships remained afloat, were constructed and the development of the steam crane greatly speeded cargo handling.

As the second half of the nineteenth century dawned, steam powered vessels had been in service for many years, but they had not challenged the supremacy of the sailing ship as a general cargo carrier. Their steam engines were simply too inefficient, which meant that they had to carry so much coal that they were unprofitable as general cargo carriers, limiting their use to prestige services such as mail routes where they could charge higher rates because of their greater speed and reliability. Hulls could be built of iron which was stronger so that ships could be made bigger and might have iron masts and wire rigging, but the advantages had to be balanced against the higher costs of construction. It required improvement in boiler technology to allow higher pressures that in turn led to the development of the triple-expansion engine, to increase the efficiency of steamships to the point where they could displace the sailing ship as the main ocean-going cargo carrier. The wooden hulled sailing ship had been dominant in 1850, but by 1900 its heyday was over. A few large sailing ships remained in service, but their numbers declined and they no longer played a major role in supporting the world's trade.

In the twenty-first century very few people have ever seen a large sailing ship let alone travelled on one. The nearest most people have come to seeing such vessels has been during the 'Tall Ships' events. However, the 'Tall Ships' as we see them today have some very significant differences to their nineteenth-century counterparts. They have radio communications, satellite navigation systems and critically, auxiliary engines that enable them to manoeuvre with a facility that could only have been dreamt about 200 years earlier. They also have life-saving equipment far superior to the simple open boat propelled by oars, which constituted the only similar equipment available to their predecessors. Nevertheless, they do convey a reasonable impression of how the sailing ship of 150 years ago would have appeared and there are many other similarities.

However, in the first half of the nineteenth century the seas they navigated were still largely devoid of aids to navigation, while captains were often unable to say with any degree of certainty precisely where their ship might be, hence the continuing rate of loss. There was general agreement that these losses might be reduced if the coasts were adequately marked, especially at night. To comprehend why lighthouses were so important and considered to be so important, it is essential to at least have a basic understanding of the way in which the cargo carrying sailing ship of the early nineteenth century was constructed and how it worked.

Construction of the Ship

The hull of almost every sailing ship built before 1850 and many built thereafter, was made from wood. However, there is a limit to the length of a piece of wood, so every part of the hull had to be made by joining pieces of wood together. The bigger the ship, the greater the number of pieces of wood that were required. The shipbuilder would start by building a sturdy frame and then cover this with wooden planks to form a watertight skin. However, this technique limited the size of a ship and the bigger the ship, the more difficult it was to make the hull sufficiently rigid. Consequently, wooden sailing ships were, of necessity, relatively small. Moreover, although a sealant known as a caulk was used, the myriad of joints between the timbers inevitably meant that a small volume of water could leak in, even under ideal conditions. In stormy weather when the sea was whipped into huge waves the strains on the hull of a heavily loaded vessel were greatly increased and the seams between the planks which clad the hull would tend to work open, letting more water into the ship. Thus, it was necessary for the crew to pump out the water every day, while in a storm the pumps might have to be in constant operation.

A ship which leaked significantly in calm waters would inevitably leak far more when exposed to the much rougher waters likely to be encountered in the open ocean and this clearly constituted a major threat to the survival of the ship and its crew. Unsurprisingly, crews were very reluctant to put out to sea in a ship that was leaking badly. In May 1870 the crew of the barque *Margaret Smith*, outward bound from Greenock on the river Clyde with a cargo of coal for Cuba, was found to be leaking so badly that her crew simply refused to leave the sheltered waters of the Firth of Clyde.[5] Consequently, she had to turn back, returning to Greenock via Lamlash. Her seams had to be re-caulked, a process that involved forcing a packing called oakum, a material made from strands of old rope mixed with tar, between the seams. In this case the *Margaret Smith* was able to resume her voyage a few days later. The same ship suffered a similar problem just over two years later in July 1872, when again on a voyage to Cuba, the crew refused to take her to sea.[6]

The vessel derived its motive power from its sails which were carried on wooden masts. The masts by themselves, which like the rest of the vessel were often in two or more pieces, could never have withstood the forces exerted by the wind on their own so they were braced in position by a complex mass of ropes for which a common general term was 'rigging.' The sails, made of canvas, were of several basic types. Sails which were more of less rectangular in shape and were suspended from a wooden beam known as a 'yard' which was horizontal and was thus at right angles to the hull were often referred to as 'square' sails and a ship which relied on this type of sail was said to be 'square-rigged.' This type of sail was very effective in harnessing the power of the wind, but presented the shipowner with a problem. To manoeuvre the ship all the sails had to be manipulated more of less simultaneously and that required a large crew.

A second common type of sail had one vertical edge connected directly to a vertical mast. The bottom of the sail was secured to a wooden beam known as a boom. The top of the sail could be attached to another timber or even to the mast. These were known as 'fore and aft sails.' A variation of the fore and aft sail were the triangular shaped sails which ran from the foremast to the front of the ship and were known as jib sails. The two types of sail were not mutually exclusive and many ships carried a combination of both types. It was the type of sails and the way in which they were arranged that gave the type of ship its name. The *Margaret Smith*, was called a barque because she had three masts, carrying square sails on the first two masts, but fore and aft sails on the third mast, the one nearest the stern which was called the mizzen mast. The drawing in Fig. 1 shows the sail arrangements of a barque. This was a very common arrangement for a cargo ship as the square sails on the foremast and mainmast provided plenty of motive power, but the fore and aft sails would provide sufficient power to manoeuvre the ship in confined waters and required a much smaller number of men.

No matter what sail arrangement a ship had, it still relied, as sailing ships had done for centuries, on the power of the wind. If the wind happened to be blowing in more or less the same direction as that you wished to take, then it was a relatively simple matter of just setting the sails to catch the wind. Having the wind in the right direction was thus extremely important and sailors took advantage of a phenomenon that had been known for many years.

Our barque, the *Margaret Smith*, spent most of her life carrying cargo between the UK and the West Indies, usually coal on the outward voyage and sugar on the inward. To make the voyage as quickly as possible the captain would make use of a phenomenon well known to all sailors, the prevailing winds. It had long been known to sailors that there was a basic pattern to the way in which the winds blew and they could make use of this information to speed their voyages. In Fig. 2 there is a very simple representation of the prevailing wind pattern that the *Margaret Smith* could use. On the outward voyage she would sail well to the south to catch what were known as the 'Trade Winds' which would carry her to the West Indies. On the homeward voyage she would sail to the north to make use of the 'Westerlies.'

Fig. 1   Barque.

Fig. 2   Basic Wind Pattern for the North Atlantic.

The prevailing westerly winds would propel the ship effectively back to Britain, but accurate navigation was essential to ensure that after the long trans-Atlantic journey the ship was correctly positioned to sail between the south of Ireland and northern France.

In light winds the ship would need every possible sail to catch as much of the wind as possible, but as the wind grew stronger the force it exerted on the sails created ever greater strains on the masts and their supporting rigging. Unless the area of sail exposed to the strengthening wind was reduced, this strain would eventually become too great, the rigging would no longer be able to provide sufficient support and a mast could break with disastrous consequences for the ship and her crew. Consequently, when the wind blew strongly as it did in a storm, the crew were forced to reduce the area of the sails, either by reducing their size or by fastening them to the yards or masts so that they no longer caught the wind. Storms in the northern Atlantic are commonplace, especially in the winter months and the *Margaret Smith* would experience her share of them. In 1872 she encountered a severe storm on a voyage from Matanzas in Cuba to Greenock. A heavy sea washed one sailor, William Reid, overboard and did severe damage to her deck fittings, while also damaging her masts, rigging and sails.[7] This damage was noted by a another passing vessel, the *Howard D Troop*, who reported the *Margaret Smith*'s plight.[8] She endeavoured to seek shelter in Cork, but the strong winds proved too much for her to manage in her damaged condition and she was forced to continue east, finally finding shelter in Waterford.[9] Here it was decided that the best course of action would be to patch her up sufficiently to make the short crossing of the Irish Sea to Bristol where her cargo could be unloaded.[10] Once her cargo had been discharged she had to be towed back to the Clyde where she was repaired in an Ardrossan shipyard, work which took many weeks.[11]

It might be imagined that a simple solution would be to just stop and wait for the storm to pass, but that would mean anchoring the ship. However, it is only possible to anchor a ship in relatively shallow water and the ocean is deep, often very deep and hence anchoring was impossible. Therefore, there was no alternative but to sail on, albeit under reduced sail. Reducing the amount of sail might be essential, but unless the ship retained some motive power from her sails it was impossible to keep the vessel on course and that could very quickly lead to disaster. Ships of all types were and still are designed to cope with waves that meet the front, the bow, or meet the rear, the stern. This means that the smallest possible area of the hull meets the wave and while it may make the ship pitch up and down, it not likely to do much damage and the ship should survive. However, if the ship is allowed to turn so that the side of the ship is exposed to the waves, then the force of the wave will be trying to push the ship over. At best the ship will lean or heel over to a degree which would make it very difficult to move about. However, if the ship rolled too much, then the deck on the side away from the waves, the lee side, could be forced under the water with the very real danger that water could find its way into the hull. This would have the effect of making the roll greater until finally the ship was lying on its side. In this position it would be almost inevitable that water would enter the hull in large quantities with the result that the ship could roll right over or capsize. A sailing ship was particularly vulnerable to wind and waves from the side as the effect of the wind on the tall masts and rigging exerted a powerful force trying to roll the ship over.

The consequence was that it was at best extremely risky and often impossible to turn a sailing ship around in a bad storm. This could lead to a heart-breaking situation such as the one experienced by the crew of the *Margaret Smith* in March 1867. On a voyage from Demerara to Liverpool she encountered such severe stormy weather that to try to even change course would have been almost suicidal. Thus, when they sighted a British barque in distress they were unable to go to her assistance, or even get close enough to discover her name. The crew of this vessel had been forced to take to the rigging, but nothing could be done to help them.[12] The men of the *Margaret Smith* could only look on helplessly, knowing that these men, their fellow seamen, were doomed.

In strong winds it would be essential to reduce the area of sail to prevent damage to the masts and rigging. There were two ways of doing this. If the reduction required was small then the sails could be partially furled, that is rolled up so that less of the sail was exposed. This was referred to as reefing a sail and some vessels were equipped with patent self-reefing topsails that could be reefed very quickly without the need for sailors to go aloft.[13] If further reduction was required, then the sail could be completely furled so that none of it was exposed.

Fig. 3   A barque with reefed topsails.

It must have been a daunting task to climb the rigging of a ship in even a modest storm. It would be vital to have a secure handhold to prevent the wind blowing you off the rigging, while at the same time the ship was rolling and pitching in heavy seas, a motion that would be felt more strongly the higher you climbed. It was certainly a task that called for skilled seamen. However, as we shall see, reducing the area of sail to prevent damage might be a necessity in a storm, but it also presented a problem.

Now it is fairly easy to see how a ship might sail in the direction of the wind, for we are all familiar with the fact that the wind tends to move things in the direction in which it is blowing. We are also familiar with the difficulty of trying to move in the opposite direction, against the wind. We tend to bend forward and lean into the wind as we force one leg in front of the other. Here was a major problem for any sailing ship, how could you sail against the wind? In ancient times one solution was to abandon the sails and simply use rows of oars to provide the necessary power. It was a viable way to move a vessel forward, but did require a lot of men to man the oars. That might be fine if you could get the rowers to work for nothing, but hopelessly uneconomic if you had to pay them. Gradually, sailors evolved techniques that would allow them to make progress against the wind, albeit that they could not sail directly into the wind. However, as it was the problem of sailing against the wind that was a major factor in creating the demand for lighthouses, we need to have a basic understanding of the way in which it was achieved.

No sailing ship can sail directly against the wind, but with the correct technique it is possible to sail at an angle to the wind. This is illustrated in Fig. 4 where the ship is steering a course which allows it to move forward while at the same time moving to the side. In order to achieve this the sails have to be aligned so that they are at an angle to the direction of the wind. In order to understand how this enables the ship to move forwards we need to consider a bit of basic physics. When the wind pushes against the sail of a ship it exerts a force on that sail and that force can be divided into two components, rather as if there were two separate forces, one component X pushing the ship ahead through the water and the other component Y at right angles to component X which is trying to push the ship sideways through the water. This is illustrated in Fig. 4. In this case the ship is seeking to move in a westerly direction but is also moving to the south.

Fig. 4   Sailing against the wind.

The wind is blowing from the West and creates a force F on the sail.

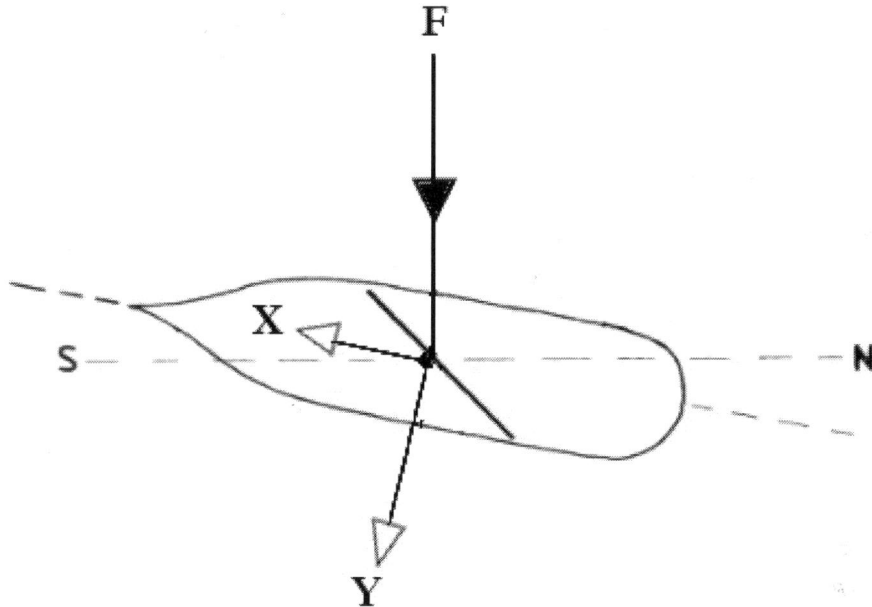

There are several points about this diagram that need to be considered. The first is that both components X and Y are each only a proportion of the original force exerted by the wind and the second point is that they are not necessarily equal. Their respective magnitude depends on the angle at which the wind strikes the sail. However, as it takes far less force to move the ship forward than sideways, the result is that even though component X may be smaller than component Y, the ship moves forward more than it moves sideways. In reality, component Y does move the ship sideways to some extent so the actual path the ship follows is not the one in which the ship is pointing, but in this instance one nearer to the south.

In Fig. 4 the ship is moving in a direction to the west of south. It is gaining ground to the west which is what is desired, but is also moving to the south. At some point the ship needs to turn to the right, to starboard, where it can again achieve some westward movement, but this time also moving to the north. In other words, the ship follows a zig-zag course, but slowly makes progress in the desired westerly direction. The process, known as tacking, enables the ship to move forward against the wind, albeit at the expense of having to sail many more miles in the process. Sailing against the wind was something that sailors sought to avoid if it was at all possible and it was usually preferable to choose a course which might involve a greater distance if it meant that the wind was blowing in roughly the same direction as the ship needed to travel.

In moderate weather the process is perfectly viable but if the strength of the wind rises the ship begins to encounter a problem. It was noted earlier that in very strong winds it would be necessary to reduce the area of the sails to avoid breaking the masts and the rigging. This creates a real dilemma when trying to tack in a storm. The force which actually drives the ship forward was, as we saw in Fig. 4, component X. However, the magnitude of component X depends on the area of sail and so if the area of sail is reduced, so is the magnitude of component X. In other words, the force moving the ship forwards is smaller, while the wind is still pressing on the masts and rigging without sails, forcing the ship sideways. This gives rise to two problems for the sailor. The first is that the point will be reached when although the ship is pointing forward and actually moving forward through the water, the sideways movement is so big that the ship is actually making no progress in the desired direction and may even be being driven nearer the shore as shown in Fig. 5. The second is that while the ship is turning from one tack to the other the sails are providing no driving force at all and it is only the momentum that the ship has built up that enables it to make the turn. If the reduced driving force is not sufficient to give the ship sufficient speed through the water, then it may not be able to complete the turn. If that happens then the ship could easily become helpless and out of control with potentially disastrous consequences.

In Fig. 5 the ship has sailed between two headlands and is now effectively trapped. The storm has forced the captain to reduce sail but now there is insufficient force created by the sails to drive the ship on the course set and it is being forced relentlessly backwards. Once in this position a vessel was virtually doomed. If there had been a lighthouse on the headland as shown and another on the other side of the bay, that is about 30 nautical miles apart, then there is every likelihood that the captain would have had sufficient warning to take evasive action.

The point of all the foregoing is to explain just how dangerous it was for a sailing ship to approach land, especially in stormy weather. It created a dangerous situation but one which could not be avoided. In order to make reasonable progress the ship needed to have her stern to the wind so that her sails would efficiently drive her forward, but at the same time because it was difficult and often impossible to stop or reverse course, it was essential that the captain did not find his ship approaching land unexpectedly. It was, therefore, very important that he knew exactly where his ship was which raises another crucial issue which is discussed in the next chapter.

Fig. 5   Driven to Lee.

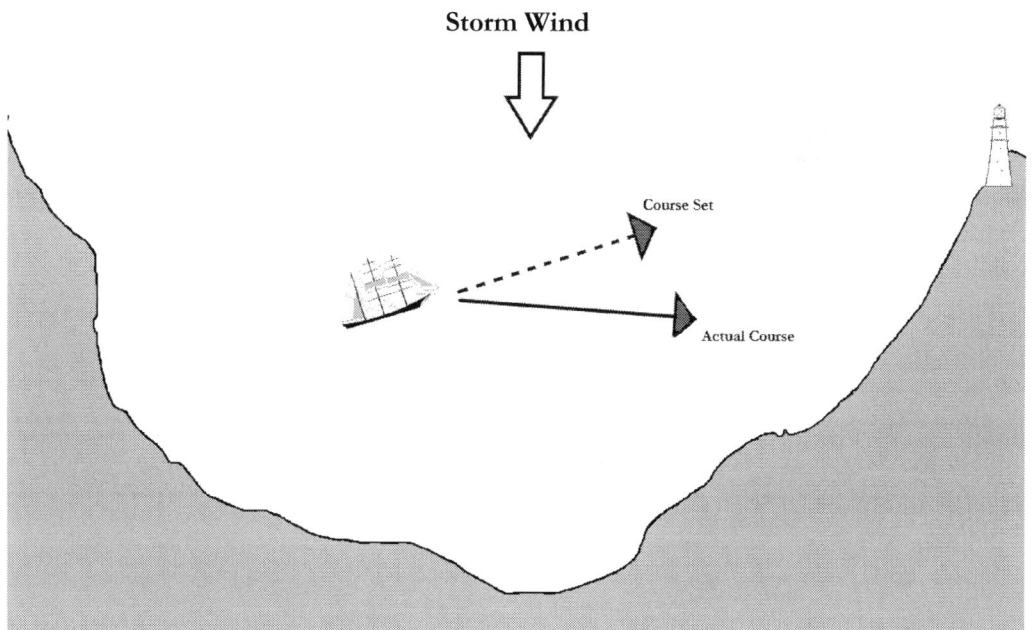

[5] Lloyd's List 24th May 1870.
[6] Lloyd's List 11th July 1872.
[7] Greenock Telegraph 13th December 1872.
[8] Greenock Telegraph 11th December 1872.
[9] Lloyd's List 12th December 1872.
[10] Lloyd's List 18th December 1872.
[11] Greenock Advertiser 24th April 1873.
[12] Lloyd's List 16th April 1867.
[13] Greenock Advertiser 6th February 1857.

Chapter Three
The Trackless Ocean

Anyone who has ever made an ocean voyage will be well aware that one bit of the ocean looks just like every other bit. This naturally made it very easy for the first sailors to get lost and consequently they tended to sail along the coast, always keeping the land in sight. However, this could lead to voyages being much longer than necessary, particularly if the sailors knew that their destination was in a particular direction. If they could ensure that they always sailed in the correct direction then they could be reasonably confident that they would arrive at, or close to, their destination. It had long been known that this could be achieved by means of a magnetic compass and this became one of the first important aids to navigation. All they needed was some kind of simple map which could be used to work out the direction in which they needed to sail and then by always keeping the compass needle pointing in the chosen direction they could ensure that they were sailing in that direction. A basic compass is illustrated in Fig. 6.

Fig. 6   Magnetic Compass.

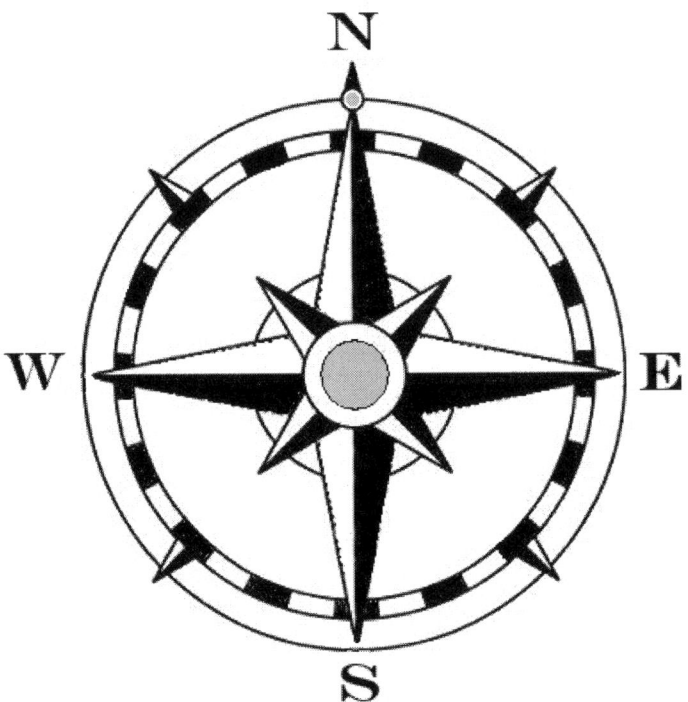

The magnetic compass was a simple and reliable piece of equipment that mariners had used for many years. It was generally mounted in a binnacle[14] near the ship's steering wheel which allowed it to remain steady and level no matter how the ship rolled or pitched. In a wooden hulled ship like the *Margaret Smith*, there was nothing to distort the Earth's magnetic field and as long as no large iron object was brought close to it, then the magnetic compass was a reliable instrument. The introduction of iron hulled ships would, however, make the use of the magnetic compass much more complex as the iron hull distorted the earth's magnetic field. The dangers this created were graphically exemplified by the loss of the RMS *Tayleur*. She was a large iron hulled sailing ship, built in Warrington in 1853 and owned by the White Star Line, a prestigious vessel designed to carry passengers to Australia. Incorporating many new technological advances such as a sub-divided hull, it had originally been envisaged that she would have two auxiliary steam engines, but part way through construction the design was changed to that of a fully sail powered vessel, a change which required re-positioning of the masts.[15]

Prior to her maiden voyage the *Tayleur* had, as was the usual practice in those days, undergone no sea trials and her master, Captain John Noble, had expressed concerns regarding the accuracy of her compasses. The *Tayleur* had three compasses installed and adjusted by John Gray, the leading compass expert for the area. He had adjusted them and declared them to be accurate, but Captain Noble remained dubious and tragically, his concerns turned out to be well founded. The *Tayleur* sailed from Liverpool on the 19th January 1854, intending to sail down the Irish Sea and then west out into the Atlantic before turning south. However, while his compasses indicated that the ship was on a course that would take her safely down the Irish Sea, she was instead headed for the coast of Ireland to the west. On the night of the 20th/21st January she ran into the cliffs of Lambay Island in Dublin Bay and well over half of the 660 persons aboard perished.[16] The pride of the White Star Line, a vessel hailed as one of the largest and finest afloat had been lost on her maiden voyage, an uncanny prequel to the more famous disaster of 1912 which befell the same company's RMS *Titanic*.

The second requirement was a map which showed the various places to which the ship might want to sail. Maps had been known since ancient times so would have been a familiar item for the sailor, but they were still essentially maps which relied upon the land to enable the sailor to use them to find their way. Over the years these maps became specialized for the use of mariners for they started to include features of land that was invisible, the land beneath the ship. The sea is not the same depth everywhere, so if the depth had been measured in a particular place that could be marked on the map. Similarly, if the sea-bed was made of sand or mud, that too could be marked. These specialized maps for the mariner were called charts. In coastal waters they were very helpful because there was almost always some reference point either on land. or the sea-bed, which would enable the ship's position to be established. Fig 7 shows a nineteenth century chart which covered the approaches to the Indian port of Bombay (Mumbai).

The chart provides a lot of detailed information about the waters close to the shore where it is shallower. However, once the water becomes deeper as it does as you head west out into the Indian Ocean, then the chart is simply blank, hence in the deep waters of the ocean the chart would offer far fewer clues as to the ship's position as there was no obvious reference point.

Fig. 7   Chart of the Approaches to Bombay (Mumbai)

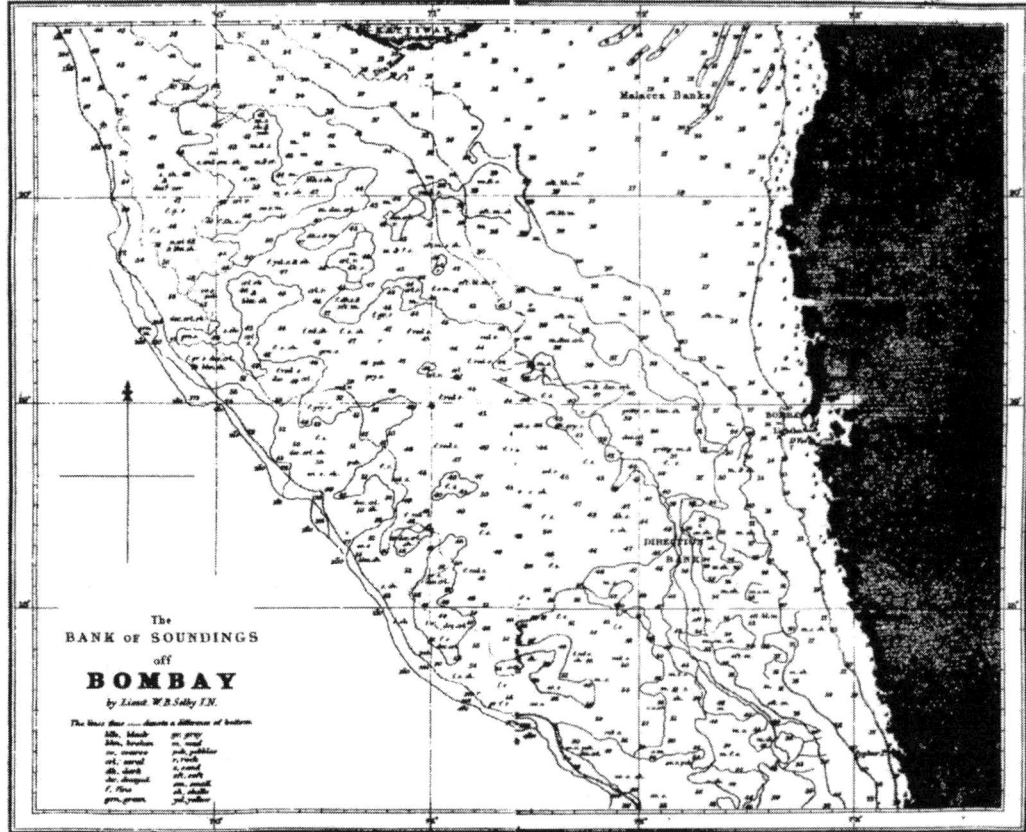

Navigation in the Oceans

Trans-oceanic navigation required a very different set of techniques. While it is not necessary to fully understand these techniques, it is important to understand a little about the basics of navigation. It had been known for centuries that the Earth was essentially a sphere and to help define any position on the surface of the earth, two sets of imaginary lines were invented to create a rectangular grid. The horizontal lines circling the Earth are called lines of latitude and are numbered from the line defining the middle of the Earth which is called the Equator.

The vertical lines which run from the North Pole to the South Pole are called lines of longitude and again they are numbered. There was no obvious basis for the numbering as there was with the lines of latitude. However, because much of the basic work was done in Greenwich, London it was decided that this place would be used as the key reference point. Hence the line of longitude that runs through Greenwich was used as the baseline and all the other lines of longitude were numbered according to the distance from Greenwich. The basic idea is shown in Fig. 8.

Fig. 8   Earth with Latitude and Longitude

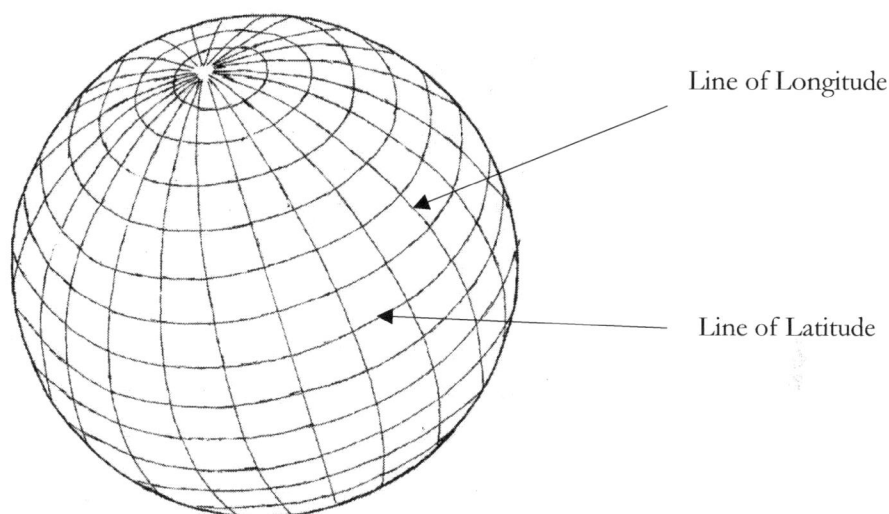

It will be obvious from Fig. 8 that the effect is to divide the Earth into a series of spaces approximately square in shape. They are not precisely square as the lines of longitude get together as they near the poles, but for our purposes we shall consider them as squares. Any position, even one in the middle of the ocean could be defined in terms of latitude and longitude, but that meant having some means of determining your position in relation to latitude and longitude.

Latitude could be determined by making use of a phenomenon with which we are all familiar. Everyone in the northern hemisphere is aware that the sun rises higher in the sky the further south you go which is why it is generally warmer in countries that are nearer to the equator, where the sun rises higher and the hours of daylight are longer. In fact, at the equator the sun would be more or less right overhead in the middle of the day. Thus, the angle between the sun and the horizon at mid-day was directly related to latitude. If you could measure the angle, then you could work out your latitude. At first, measurement of this crucial angle was made with a variety of fairly crude instruments such as quadrants or cross-staffs or the more sophisticated astrolabe. Accuracy was improved by the development of the sextant, an early version of which was developed by a Fellow of the Royal Society, John

Hadley in 1731.[17] It was extensively tested in the following year and was soon widely adopted. A simplified version of this instrument is shown in Fig. 9.

Fig. 9   Sextant

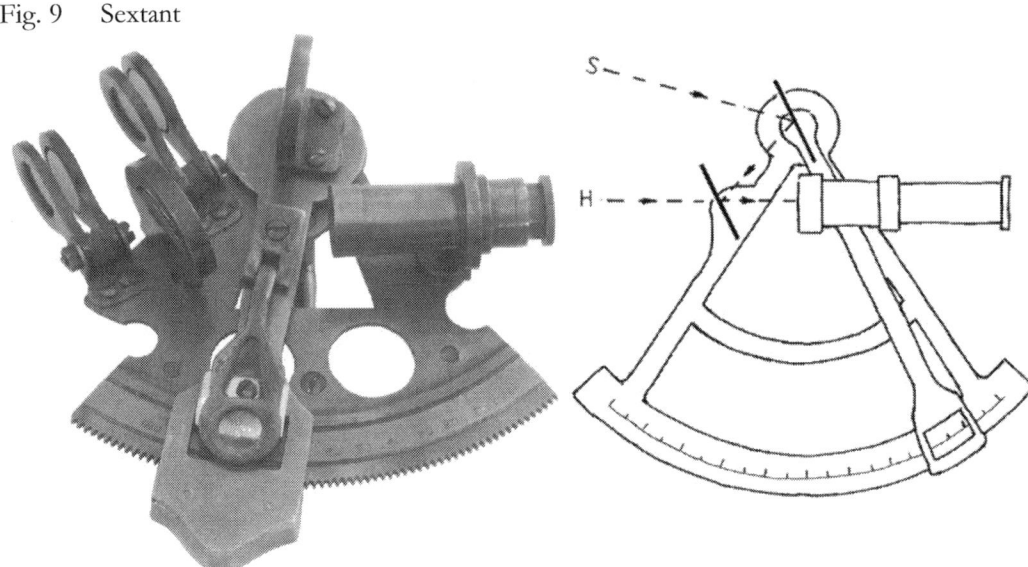

On the left is a photograph of a real sextant while on the right is a simplified diagram showing the paths of the rays of light. The additional parts shown in the upper left of the photograph are filters to reduce the intensity of the light, essential when the sun is being observed. (Photo Chris Hills)

The sextant has two mirrors, one of which is attached to a movable arm. The mariner would look through the telescope and adjust the position of the arm until an image of the sun was aligned with the horizon. The angle could then be read off from the scale. Filters of varying darkness could be used so that the user's eyes were not damaged by looking directly at the sun.

However, there is a small problem in that as it orbits the Sun, the Earth tilts on its axis with the result that for half the year the northern hemisphere is closer to the sun. The people in the north experience summer and we all know that the sun climbs higher in the summer than it does in winter. Consequently, the latitude indicated by a particular angle varies every day. It is possible to do all the necessary calculations, but there is an easier solution. Astronomers and mathematicians have produced tables which would enable the mariner to quickly determine the vessel's latitude. These tables are published in a book known as an 'almanac' where all the mariner needs to do is measure the angle between the sun and horizon at mid-day and then use the tables to find their latitude. In reality, it is a little more complex than that and the navigator has to feed various other bits of data into the calculation. A description of the method, dating from 1839 is to be found in 'The Practical Navigator' by William Black.[18] The captain of the *Margaret Smith* should have been very familiar with this book as it was published in Greenock, William Black's home town and the *Margaret Smith's* home port. Black describes the process as follows.

To Find the Latitude of a Ship by the Meridian Altitude of the Sun

Rule - To the observed altitude of the sun's lower limb add the sun's semi-diameter (16 minutes). Should the upper limb be taken subtract it; then subtract the dip of the horizon (Table IV), which will give the apparent altitude of the sun's centre; in Table VI will be found a correction for the joint effect of parallax and refraction, to be subtracted from the apparent altitude, the result giving the true altitude of the sun's centre.

By subtracting the sun's true altitude from 90° we obtain the meridian zenith distance. If the observer is north of the sun, it will be marked north, if south of the sun mark it south.

Take the declination of the sun from Table VIII, and reduce it to the meridian of the ship by Table **X**, then if the zenith distance and declination be both north or south, their sum will give the latitude of the ship; should the one be north and the other south, their difference gives the latitude of the ship with the same name with the greater.

The problem of finding the ship's latitude by means of the sun's elevation had been solved and similar methods had been developed using some of the stars. It meant that the navigator had to have a rather thick book with all the relevant tables, but it was a proven and reliable methodology. Longitude would prove to be rather more of a problem.

There was an obvious way to calculate a ship's longitude and again it relied on a phenomenon familiar to everyone. Almost everyone and certainly anyone who has travelled abroad is likely to be aware that in other countries the time is different. In mainland Europe, which is to the east of the UK, the time is ahead of UK time, while in the USA, which is to the west, it is behind the UK. This because the Earth is a sphere and spins once a day on its axis as it orbits the Sun. Therefore, the sun rises in the east and sets in the west, reaching the highest point in the sky at 12 noon. It is this highest point that defines the local time in any part of the world or indeed any part of the country. In large countries like Australia, Canada, Russia and the United States, there are agreed time zones. Until the advent of a national railway system, it was quite normal for each town, even in a small country like the UK, to operate on local time.

There is clearly a relationship between the 'time' at which the sun reaches its highest point in the sky and your position to the east or west of Greenwich. Halfway round the earth, or 180° east or west of Greenwich, the difference would be 12 hours. It is, therefore, just simple arithmetic to calculate that a time difference of an hour equates at a 15° difference in longitude. This makes it very easy to calculate your longitude because all you need to do is note the time at Greenwich when the sun is at its highest. It sounds so simple, but there was a major problem, which was knowing the time at Greenwich when you could be thousands of miles away.

The obvious requirement was for a clock which was set to the time at Greenwich and which you could carry with you. It sounds very simple today when for only a small outlay you can purchase a digital watch accurate to within a few seconds per year. Accurate clocks that would work on land were available. They made use of a phenomenon first observed by Galileo, that the length of a pendulum determines how fast it swings and the first successful pendulum clock had been built by Christiaan Huygens in 1656.[19] Subsequent clocks became more accurate and robust and experiments were carried out determine whether they would solve the problem of finding longitude at sea. They seemed to work fairly well in fair weather, but the pendulum was confounded by violent sea movements. An alternative to the pendulum was the coiled spring, an idea promoted by both Huygens and Robert Hooke, who both claimed to have invented the concept.

However, neither the pendulum nor the coiled spring mechanism offered anything near the degree of accuracy required as they were affected by issues such as changes in temperature as well as by a ship's motion. It seemed to be impossible to construct a timepiece of sufficient accuracy and an alternative method was sought. The lines of longitude divided the Earth into 360 degrees and at the equator where the Earth's diameter is greatest, 1 degree was equivalent to 60 nautical miles or about 68 geographical miles. If we imagine that the best timepiece available had an accuracy of 99.9% then the error over each 24 hours would be about 86 seconds or almost 1½ minutes. Now one minute of time was equivalent to 0.25 degrees so after only one day, the 99.9% accurate timepiece would be inaccurate to the extent of 0.375 degrees. Moreover, that error would increase each day and given that a voyage across the Atlantic could take as much as two months, then the error at the end of that time could be as much as 22.5 degrees or well over a thousand nautical miles. The level of accuracy required needed to be much higher, in fact an accuracy of about 99.999%, or an error of about 1 second per day was required. Unfortunately, the technology required for such a timepiece was simply not available. The longitude of a location on land could be determined by a series of complex astronomical observations and it seemed that the only solution lay in trying to find a way in which this could be done on board a ship.

There is an old saying that 'Necessity is the Mother of Invention' and there was no doubt about the necessity. The need for a means of determining longitude was becoming ever more pressing. Far too many ships were being lost because, not being able to accurately establish their longitude, they found themselves too close to land, were driven ashore and lost. It should come as no surprise that around the coasts of Cornwall, the first land that a vessel coming from the Atlantic might encounter, there are some 6,000 wrecks.[20] In 1707 a Royal Navy fleet commanded by Sir Cloudesley Shovel had lost four ships and over a thousand men when they blundered into the Scilly Isles off the Cornish coast because they were unable to determine their longitude and believed that they were much further west. It was clear that something needed to be done.

The issue was of such importance that in 1714 Parliament passed the Longitude Act which offered a substantial reward, £20,000, about £3,850,000 today, for anyone who could devise a method of determining longitude at sea.[21] The Act specified the accuracy required, which was half a degree or 30 nautical miles.[22] The Act did not specify by what method longitude was to be determined and appointed a Board of Longitude to evaluate all submissions. The Board had authorization to provide funds to further research into any promising ideas, but these were in short supply. The difficulty in creating a timepiece of sufficient accuracy under shipboard conditions seemed unsurmountable and the only alternative was some form of astronomical method. While many strategies were proposed, the most promising appeared to be based on observations of the moon, with the result that a huge effort was made to develop a practical method based on lunar movement.

Meanwhile, one man had responded to the challenge of building a timepiece which would resolve the longitude problem. He was a Yorkshiremen, John Harrison, a carpenter and self-taught clockmaker whose long case clocks were already renowned for their good time-keeping, being accurate to within one second per month, a remarkable achievement for that period. However, a long-case clock needed a totally stable base and was thus of no use on a ship.

Harrison's target was to achieve a ship's clock accurate to within three seconds per day, an accuracy of 99.997%, even though the it would be subject to continuous motion and vibration as the ship rode the waves, not to mention significant variations in temperature. Furthermore, it would have to continue to operate even while it was being wound. In 1735 he had completed a clock which came to be called H1 and which he felt would meet the specification. It was to be tested on a voyage to Lisbon aboard HMS *Centurion* under the command of Captain Proctor. Unfortunately, Captain Proctor died suddenly shortly after his ship arrived in Lisbon and before he had time to make a report. It was left to Captain Roger Wills of HMS *Orford* to bring Harrison and H1 back to Britain, a voyage that took a month. When land was sighted, Captain Wills believed he had reached Start Point near Dartmouth, but Harrison disagreed. Using H1 to determine the ship's longitude he reckoned that the land was actually the Lizard, some 60 miles west of Start Point and it transpired that he was correct, greatly impressing Captain Wills who wrote an affidavit attesting to H1's accuracy.[23]

While Harrison's H1 undoubtedly worked, its complexity was such that nobody other than Harrison himself really understood how it worked. Moreover, he felt that further improvement was needed and consequently he was not awarded the £20,000 but was instead given a lesser sum on the condition that he produced an improved version within two years. The improved version which came to be known as H2 was ready by 1741 and successfully passed many tests, but Harrison was not satisfied with it and commenced work on a further design.[24] It took almost 20 years to complete and even then Harrison was not satisfied with it and embarked on a radically different design which he completed in 1760 and which would become known as H4. It looked like an overgrown pocket watch and achieved a standard of accuracy well in excess of that required by the original Longitude Act.

However, while Harrison had been struggling with his chronometer, the astronomers had eventually managed to devise a way of finding longitude by astronomical observation, using the movement of the moon. It was a complex process that had taken years to develop as the moon's path is complex and varies over time. It required the use of tables which had taken years of patient observation to compile, but nevertheless it worked. Unsurprisingly, astronomers were sceptical of a mechanical device which could render their work redundant and inevitably they sought to deride H4 as a 'one-off' oddity. However, in 1762 it was taken aboard HMS *Deptford* commanded by Captain Dudley Digges for a voyage to Jamaica. H4 was taken by John Harrison's son William who was obliged to set the time according to an astronomer in Portsmouth. It would then be checked by another astronomer in Jamaica. The trial showed conclusively that over a period of 81 days H4 had lost just 5 seconds.[25]

While it took a little while for the idea that a mechanical device could achieve with ease, a task which taxed highly skilled astronomers, there was no doubt that the chronometer was the way forward. It revolutionized navigation, as for the first time the mariner had a reliable and accurate way to determine his longitude. However, just because the technology was available did not mean that it was effectively utilized. Captains understood the means by which latitude was determined. They could see how the instrument to measure the angle between the sun and the horizon worked and could, if necessary, do the calculations for themselves. Thus, the accurate determination of their latitude was just a matter of utilizing their skills which had been built up over many years of experience. On the other hand, determining longitude by means of a chronometer involved reliance upon a piece of machinery whose operation was a complete mystery to them. All their previous experience of timepieces led them to doubt the almost incredible accuracy of Harrison's invention, and required them to put their trust in a device which worked in a way they could not understand. Consequently, as late as the 1840s it was commonplace for ships engaged on trans-Atlantic voyages to be navigated by latitude alone.[26]

If the mariner was able to determine his exact position, then it might be imagined that it would be a simple matter to ensure that ships never got into a position where they could be driven ashore. Unfortunately, even with the sextant, chart, compass and chronometer, there remained a problem that neither observation nor mechanical device could solve. Everything depended upon being able to make observations of the sun, but in winter especially, the sky could be cloudy for days, preventing observation of the sun and effectively rendering all the mariner's navigational equipment largely useless. However, even if the sun was not visible the captain still needed to know as accurately as possible where the ship was and the only option available was called 'dead reckoning.'

Imagine that you are in a car driving due west on a straight road at a constant 30 mph. It is simple to calculate that 3 hours later you would be 90 miles west. This is 'dead reckoning' in its simplest form, for the car will continue in a straight line at a constant speed unless the driver intervenes. However, with a ship it is not so simple. The most obvious requirement was to know the speed of the ship, but until well into the twentieth century there was no 'speedometer' on a ship. The speed of a ship through the water could be measured with a device called a 'log.' A very simple form of 'log' is shown in Fig. 10.

Fig. 10 Measuring the Ship's Speed

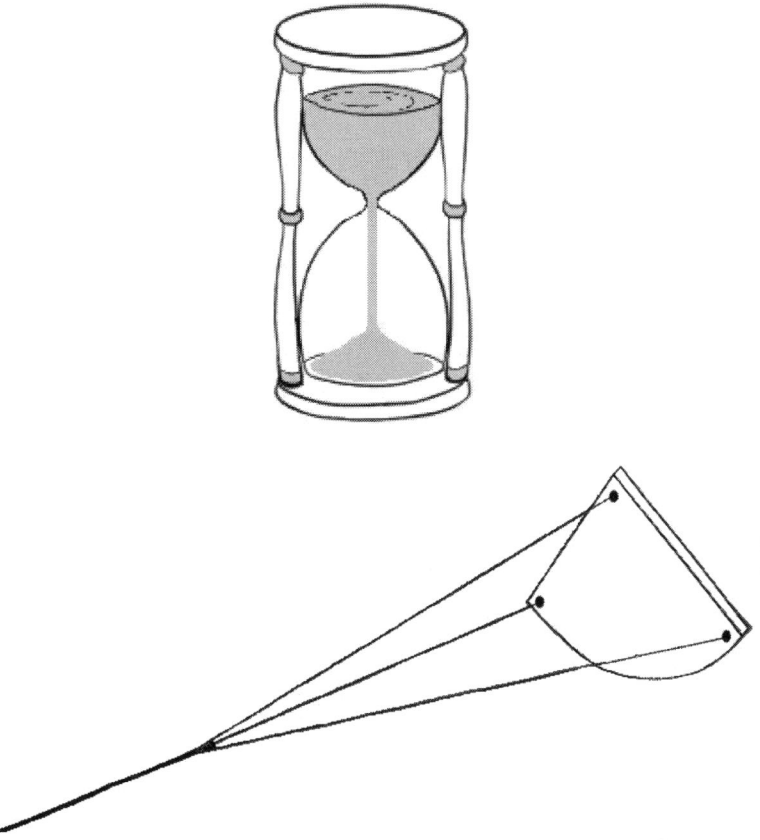

This device is just a wooden board, weighted with lead at its lower curved edge, connected by three strings to a long line so that if the line is pulled the board lies vertically in the water and will be very difficult to pull forward. If the board was thrown over the ship's side, then because the 'log' stays still and the ship moves forward, the line will be drawn out. If the length of line pulled out in a fixed period of time, as determined by an hourglass, was measured, then this would tell you how far the ship had travelled in that time and hence the ship's speed could be calculated. It was not particularly accurate and because there was not a continuous reading there was scope for error as the ship might speed up or slow down slightly according to the strength of the wind.

Another problem is that was being measured was speed 'through the water.' The road on which a car runs always stays in the same place, but the waters of the oceans are in constant motion, creating what we call currents. Thus the 'log' might indicate a speed of 2 nautical miles per hour (nmph) but if the water was moving at a speed of 1 nmph in the same direction as the ship then the actual speed of the ship would be 3 nmph in relation to the ocean floor and land, which is what really mattered. It is usual to refer to a speed of one nautical mile per hour as a speed of one knot.

Fig. 11   Major Ocean Currents in the North Atlantic

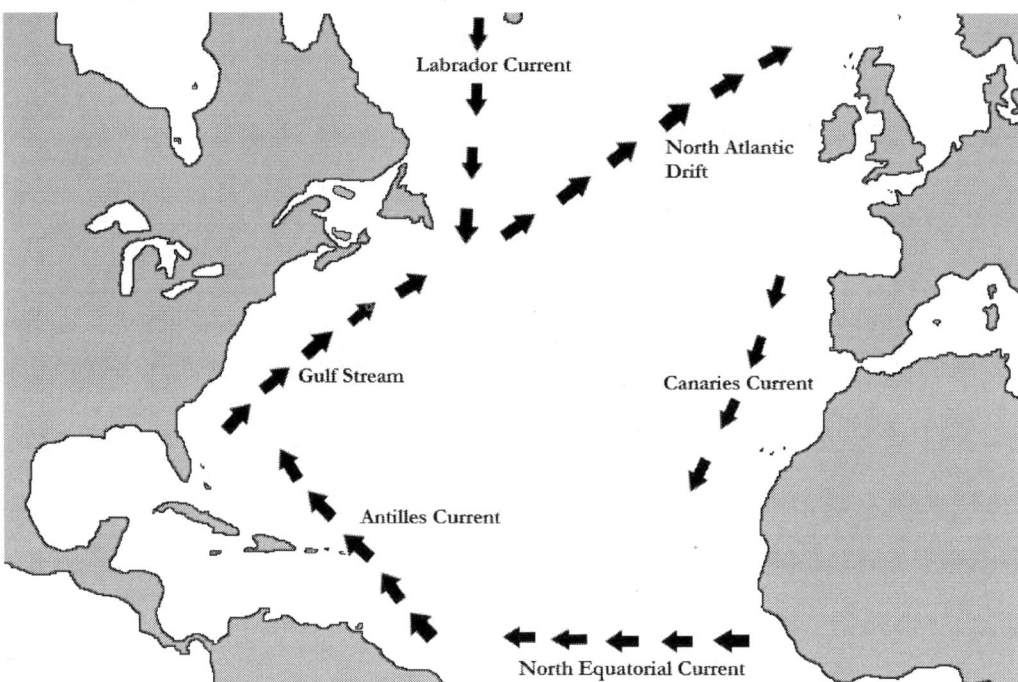

It got more difficult if the ocean current was flowing at an angle to the ship's desired course for then it would tend to push the ship sideways as well. The captain would, of necessity, have to estimate the effect of the current.

A further complication is that unless the wind was directly behind or astern of the ship, there would be a tendency for it to push the ship sideways as we saw earlier. Thus, there were a whole stack of variables which the captain could only estimate. Consequently, there was likely to be an error in the ship's position which the captain derived by dead reckoning. However, the estimated position depended upon knowing the precise starting point, but if the ship's position as derived by dead reckoning after a day was inaccurate, then the next day's position would be likely to add to the error. The longer the time for which the captain had to rely on dead reckoning, then the greater the potential error. We could easily imagine a situation where there had been no glimpse of the sun for a week. If we were to assume that the estimates of the *Margaret Smith's* skilled and experienced captain were 90% accurate, then for a ship sailing at 3 nmph the daily error would be about 7 nautical miles. Over a week that could rise to almost 50 nautical miles, more than enough to create a very dangerous situation.

The dangers of an error in position of even just 50 nautical miles are well illustrated by the map in Fig. 12 which shows the area of the sea on the approach to the British Isles. The ship must pass to the south of Ireland and the north of France which is a gap of 240 nautical miles. This considerable width meant that provided the captain could be sure of the ship's latitude, then there would be no great difficult in passing safely between them. However, the width of the gap was so large that the ship could easily pass between them without sighting land at all. That, in itself, presented no danger, but unless the captain could be sure of the vessel's longitude it would be only too easy to sail unaware onto the coast of Cornwall or its outlying islands.

However, if the ship is bound for London then it must pass to the south of Cornwall and the gap between Cornwall and France is much narrower. We shall take a further look at the potential dangers of errors in dead reckoning in a later chapter.

Fig. 12 The Western Approaches to Britain

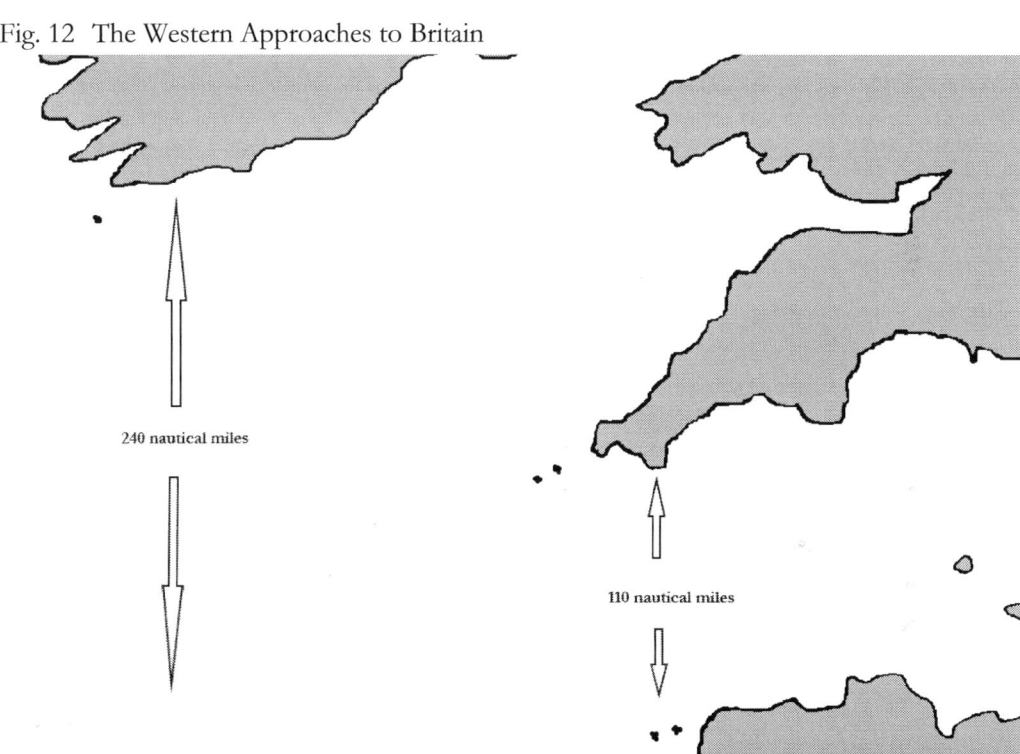

It should, therefore, come as no surprise that the coasts of Ireland, of Brittany in northern France and Cornwall were the scene of many shipwrecks. Knowledge of the ship's longitude was essential so that the perils of these rocky coasts could be avoided. Even in daylight, all too often by the time the lookout called a warning of land ahead, it was already too late as the ship would be only a few miles from disaster. However, in the dark of the night, especially if it was in stormy weather, the ship might easily be within a few hundred yards of the shore before the danger was spotted. Clearly, anything that would give an adequate warning that the ship was headed into danger would be invaluable to the mariner.

The Directory

As mariners had been navigating the oceans of the world for hundreds of years, they had learned a great deal in that time. This knowledge included information about everything from the ocean currents to the best place to anchor in a particular harbour, the position of underwater hazards miles offshore, to the height of the church tower that acted as a landmark and much more. By the early nineteenth century it had all been collected and collated into a single volume called a 'Directory'. Nowadays they are sometimes known as 'Pilots.' These directories were vitally important to the Captain of a ship, especially if they were approaching a shore with which they were unfamiliar. The frontispiece of a nineteenth century Directory is shown in Fig. 13.

Fig. 13   Frontispiece for the Alexander George Findlay's Directory for the Indian Ocean 1866

Alexander George Findlay had been producing these 'directories' for some time and he would undoubtedly have been aware of the importance of accurately recording the position of every possible thing that would aid navigation. He would know only too well that the lives of sailors could easily depend upon the reliability of his work as well as the necessity of ensuring that the information was as up-to-date as possible. Keeping the information in the 'directories' up-to-date was major task as these were substantial volumes. Findlay's Indian Ocean Directory of 1866, a volume which will play an important role in a later part of this book, ran to some 1,142 pages. In 1860 the *Margaret Smith* had undertaken a voyage from Greenock to Bombay (now Mumbai) under the command of Captain McNab who might well have found the comprehensive guidance in the directory of considerable value. A sample of this guidance is reproduced in Fig. 14 below.

Fig. 14 The first sections of the Guidance offered for vessels approaching the port of Bombay.

Directions.

APPROACHING BOMBAY.—In a former part of this work we have given those general directions necessary for traversing the Indian Ocean towards that port. The bank of soundings off it has been surveyed by Captain W. B. Selby, I.N., and illustrating his chart, he has given the following useful remarks.

That very desirable object to which my attention was particularly drawn, viz., the practicability of the navigator ascertaining his position when running from Bombay in thick weather, has, I think, been completely attained; and though a careful study of the chart will be a sufficient guide, I beg to offer a few remarks, and draw attention to one or two points, as more particularly deserving careful attention.

In coming from the westward, and being in lat. 18° 30′ N., a vessel will first strike soundings of 55 fathoms coarse sand, decayed coral, and broken shells, in long. 70° 30′ E. Should, however, the latitude be doubtful, with tolerably correct longitude, that point will be at once ascertained, as, on *this* meridian, the depths will decrease to 51 fathoms fine sand, and even 48 fathoms fine sand, as far North as lat 19° 16′, while 10 miles to the southward no bottom at 300 fathoms will be found.

Again, should the latitude be correct, and longitude doubtful, 4 miles West of this meridian, no bottom would be had at 300 fathoms, while 20 miles to the eastward, 47 or 48 fathoms fine red sand would be obtained.

It is worth noting that the guidance is very detailed and refers not only to the depth of the water but the nature of the sea-bed. The text in Fig 14 above, is around half a page of the four and a half pages of guidance offered by Findlay's Directory. It was clearly written by a very experienced and skilful mariner, in this case Captain B. Selby of the Indian Navy. As the person who collated all this information, Alexander George Findlay would have readily understood the significance of knowing your precise location and what were the key features that mariners found particularly useful. He could not have been unaware that lighthouses were probably the most significant of those key features.

The obvious way to give the sailor a warning about a dangerous coast ahead would be to put a light on it because that could be seen at night and if the light was placed at the top of a tower, then the tower would be visible during the day. However, if the solution to this obvious need was so simple, then why were there so few lighthouses in the first quarter of the nineteenth century? The answer to that question will be investigated in the next chapter.

[14] A binnacle was a column-like structure fastened to the deck near the ship's wheel and where it could be seen by the helmsman. The compass was mounted inside the top of the binnacle on pivots which allowed the compass to stay level despite the movement of the ship.
[15] The Sinking of RMS Tayleur – Gill Hoffs, Pen and Sword, Barnsley, 2015.
[16] The Sinking of RMS Tayleur – Gill Hoffs, Pen and Sword, Barnsley, 2015.
[17] Sextant – David Barrie, William Collins, London, 2014.
[18] The Practical Navigator – William Black, Greenock, 1839.
[19] Longitude – Dava Sobel, Harper Perennial, London 2011.
[20] The Shipping Forecast – Nic Compton, BBC Books, London 2016.
[21] Bank of England Inflation Calculator. https://www.bankofengland.co.uk/monetary-policy/inflation/inflation-calculator
[22] Longitude – Dava Sobel, Harper Perennial, London 2011.
[23] Longitude – Dava Sobel, Harper Perennial, London 2011.
[24] Longitude – Dava Sobel, Harper Perennial, London 2011.
[25] Longitude – Dava Sobel, Harper Perennial, London 2011.
[26] The Ship, The Life and Death of the Merchant Sailing Ship – Basil Greenhill, HMSO, London, 1980.

Chapter Four
The Unlit Coast

There were several factors inhibiting the effective lighting of the coasts. The significant costs involved in erecting the necessary structures, the difficulties in ensuring that there were enough trained and reliable men to operate the lights and different views as to where lights should be sited all inhibited the lighting of the coasts. However, even if the political will to achieve an adequate standard of lighting had existed and been translated into the provision of adequate finance, there would still have been one major difficulty. The technological means to provide adequate illumination was simply not available. Not until several critical developments had made it possible to create a viable aid to navigation did the move towards providing a comprehensive lighting system gather real momentum. It will, therefore, be necessary to at least provide an outline of the problems and the technological solutions by which they were finally overcome.

The need for adequate lighting of the coasts had been obvious ever since humans first navigated the oceans and we know from history that there had been some very early attempts to meet that need. The most famous of these was the Pharos of Alexandria in Egypt, a huge tower erected around 280BC. On top of the tower a fire was kept burning, the light from the flames providing guidance for sailors. However, anyone who has ever tended a bonfire will be only too well aware that it requires a constant supply of fuel to achieve a lot of flame. If the fuel is not continuously supplied than the fire quickly dies down to a red glow which is of little use in terms of generating light. The problem with a lighthouse like the Pharos is obvious, it requires a very large amount of fuel and the labour of men to carry the fuel up the tower and to look after the fire, all of which tends to be expensive. Moreover, there is another difficulty that no amount of money would resolve. In stormy weather, when the light is most needed, the strong wind will tend to blow the flames horizontally, effectively making it far less visible to those out at sea. However, despite the inefficiency of using an open fire as a light source, the lack of an effective alternative meant that they had to be employed, albeit only in small numbers and at critical locations.

The open fire, usually using coal rather than wood as a fuel, was not completely replaced in lighting the coasts of the British Isles until 1823. Amongst the last lights using the open fire were those on the Isle of May marking the entrance to the Firth of Forth where there was a coal fire until 1810, Flat Holm in the Bristol Channel until 1820 and St Bees Head on the Cumbrian coast which retained its coal fire until 1823.

As light could also be obtained simply and cheaply from lamps and lanterns of various kinds which had been available for centuries, it might seem curious that they had not been employed to light the coasts until the nineteenth century. In fact, they had played a role in navigation, marking the entrance to harbours and some marine hazards, where they were effective as long as the distance over which they were expected to operate was quite small.

However, they had a fundamental weakness for use as navigation aids, they simply did not produce enough light. Until the advent of electricity, light came from a flame and if you wanted more light then the only way to get it was to use either a bigger flame or multiple flames, which raises many of the same issues of cost and fuel consumption that limited the use of open fires.

To understand the problem, we need to look at the way in which light was produced by a flame. Whether you used a lamp or a candle, the basic principle was the same. A wick made from some material that had a fibrous structure was arranged with one end in a liquid fuel. The liquid would rise up the wick to the top by capillary action and here it was ignited. The purpose of the wick was to allow the air, essential for the combustion process, to mix with the fuel and consequently wicks were, of necessity, in the form of either relatively thin strings or narrow ribbons. They had to be in this form as the fuel and air could mix only at the surface of the wick, so they had to present the maximum surface to the air. This inevitably limited the size of the flame and hence the quantity of light produced. If you needed more light, then you used more flames, hence the candelabra used to hold multiple candles in large rooms.

The light from a flame travels out in all directions, much as if the light was at the centre of a large, transparent ball. However, the only light that you see is the very tiny fraction of all that light which enters your eye, the rest achieves no useful purpose as far as you, in this case as the sailor out at sea, are concerned. The mathematics illustrates the problem.

If we assume that our candle flame is 1 kilometre away, then it is effectively at the centre of a transparent ball with a diameter of 2 kilometres. Light from the candle will illuminate the whole of the surface of that ball, a total surface area of 12,560,000,000 square metres, a huge number. Yet the pupil of your eye is only a tiny fraction of 1 square metre, so the amount of light entering your eye from a single candle flame just a kilometre away is so small as to make it virtually invisible. When you consider that in order to be a useful aid to navigation the mariner needs to see the light at a distance of 20 kilometres, then it is easy to see why a light which was dependent upon a few candles would have been of limited value.

If we consider Fig. 15, then it is obvious that as far as the mariner is concerned, most of the light from the candle flame is wasted because it is going in directions that would never reach their eye. One obvious way to increase the amount of useful light would be to re-direct some of the light that was going in the wrong direction, shown as ray A in the Fig. 15, and the simplest way to do that would be to use a mirror. Everyone is familiar with the fact that a mirror will reflect light and the basic principle of reflection is illustrated in Fig. 16.

Fig. 15   Candle and Sphere model

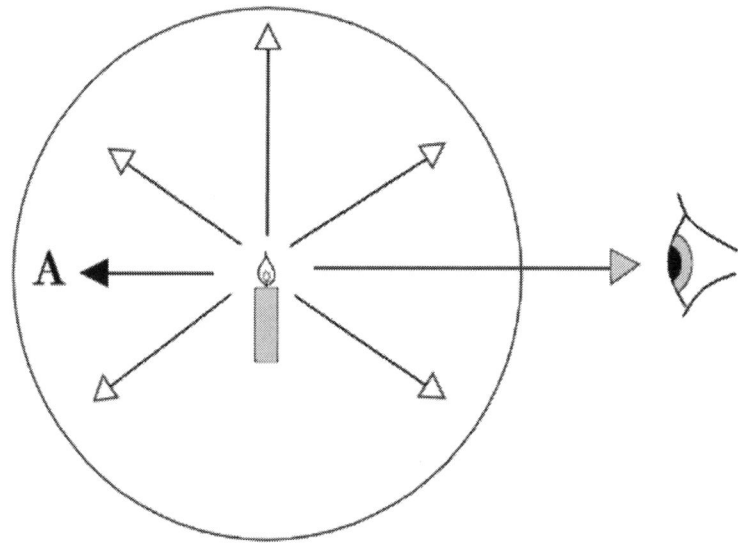

Fig. 16 The Principle of Reflection

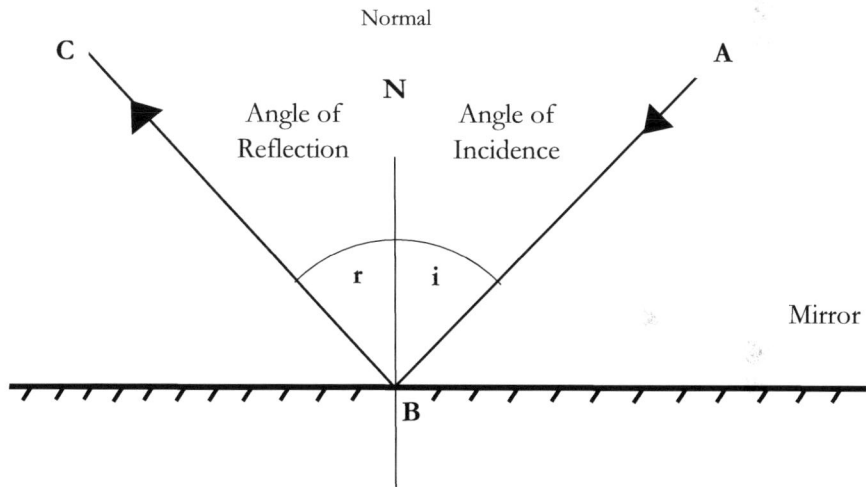

In Fig. 16 a ray of light from point A, travelling in the direction shown by the arrow, strikes the mirror at point B. It is reflected from the mirror towards point C. If an imaginary line NB, known as the 'normal', is drawn at right angles to the surface of the mirror, then the incoming ray of light, called the incident ray, will make an angle **i** with the normal. This angle is called the angle of incidence. Similarly, the ray from B to C, called the reflected ray, also makes an angle **r** with the normal. It is a basic law of physics that these two angles are always the same.

Angle of Incidence = Angle of Reflection

Now let us see what would happen if we arranged a series of mirrors behind the candle flame. This is illustrated in Fig. 17.

Fig. 17 Reflection using a Series of Mirrors at Different Angles

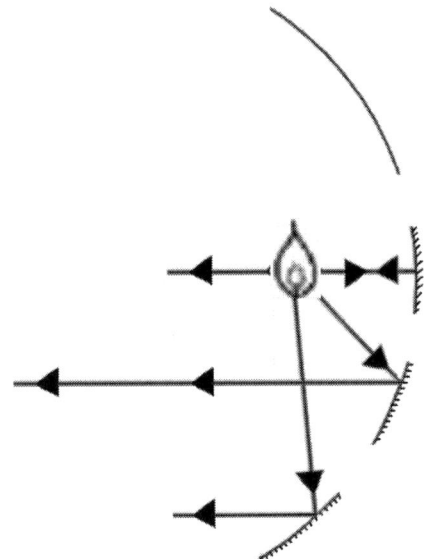

In order to reflect the light in the same direction, each mirror has to be at a slightly different angle. If enough tiny mirrors were used to reflect every ray of light, then they would form a curve, the shape of which is called a parabola. In this way a substantial proportion of the light from the candle flame could be directed in the most useful direction and hence the sailor would see a much brighter light which meant he could see it at a greater distance. In the front of his first List of Lighthouses published in 1861, Findlay gives us a fascinating story of this development in lighting technology.[27]

The first parabolic reflectors for lighthouses were used at Liverpool, probably in 1763, certainly before 1777, for in that year William Hutchinson, Dock Master of that place, published his 'Practical Seamanship,' and in that work he fully describes the apparatus used in the four lighthouses built at Liverpool in 1763.

The origin of their use is curious. It is said, that at a convivial meeting of some scientific men in Liverpool prior to this date, that one of the company wagered that he would read a newspaper at a distance of 200 feet by the light of a farthing candle. This he afterwards won by means of a wooden bowl, lined with putty, in which facets of looking glass were embedded and formed a reflector. One member of the company was William Hutchinson who seized on the idea and utilized it.

These reflectors were formed to a parabolic curve by a somewhat rude process which he describes.

> "We have had, and used here in Liverpool reflectors of 1, 2 and 3 feet focus and 3, 5¼, 7½, and 12 feet diameter. The smallest made of tin plates soldered together, and the largest of wood covered with plates of looking-glass, and a copper lamp, the cistern part for the oil and wick stands behind the reflector, so that nothing stands before the reflector to interrupt the blaze of the lamp acting upon it, but the tube that goes through it with a spreading burner mouth-piece, to spread the blaze parallel thereto, and with the middle of it just in the focus or burning point of the reflector."
>
> "The lamps are like the reflectors, proportional to make a greater or less blaze as required, their burning parts are from 3 to 12 inches broad, and are trimmed every four hours."
>
> "Thus are these lighthouses constructed, kept and situated, and have stood the test of a fair trial, and the preference and advantage given to them even by their opponents, as there will always be to new things commonly calling them new whims, till time and trial confirm them as useful improvements."

Hutchinson's description very clearly illustrates the somewhat primitive nature of such lights, but they undoubtedly represented a major step forward. It should be noted that in addition to the reflector, the size of the flame and hence the quantity of light has also been increased by the use of a wider burner. However, the wider the burner the less efficient it becomes as a light source as only the centre of the flame is correctly aligned with the mirror. This would mean that much of the light from Hutchinson's 12 inch burner was wasted. A more compact source of light was needed.

The problem of a compact light source was solved in 1780 by a Swiss scientist, Aimé Argand. He invented what we now call the Argand burner which was patented in England in 1784. The burner used a circular wick held between two concentric metal cylinders. Air could pass through the hollow central area of the cylinders so that both sides of the circular wick received a good supply of oxygen and thus burned efficiently and brightly. As Fig. 18 shows, it was like having a mass of candle wicks arranged in a compact circle.

Fig. 18   The Principle of the Argand Burner

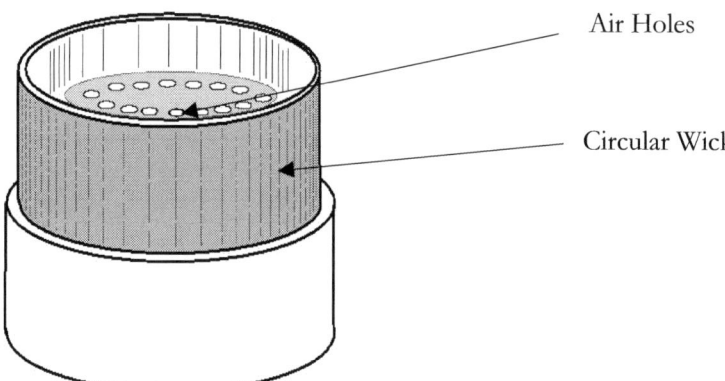

The flow of air was further increased by the use of a glass chimney and the outcome was a lamp that was some ten times brighter and burned more cleanly, as with the enhanced air flow combustion was more complete. There was an additional advantage in that the wick needed far less frequent trimming. While the Argand lamp was a great advance in lighting generally, it offered precisely the kind of intense, concentrated light source that was needed for an effective navigational aid.

Improvements in the construction of the parabolic mirrors increased the efficiency of lamps while the Argand burner with its glass chimney to channel combustion products away, allowed the reflector lamp to become steadily more efficient. The lamp could be placed at the focus of a much larger parabolic reflector as shown in Fig. 19.

Fig. 19   A Reflector Lamp with an Argand Burner

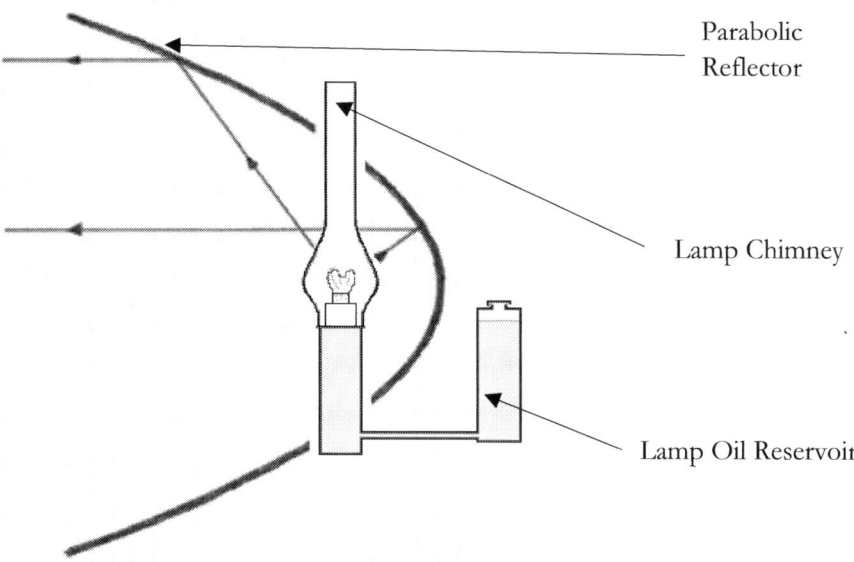

However, while the reflector lamp as shown in Fig. 19 had many advantages, it also had a major drawback. The light was concentrated in a specific direction which meant that it could only be seen from a limited arc in front of the lamp. Consequently, a lighthouse would need a number of lamps to cover every sector of the horizon. This is illustrated in Fig. 20.

Fig. 20 A set of Seven Reflector Lamps arranged to cover 180° of arc.

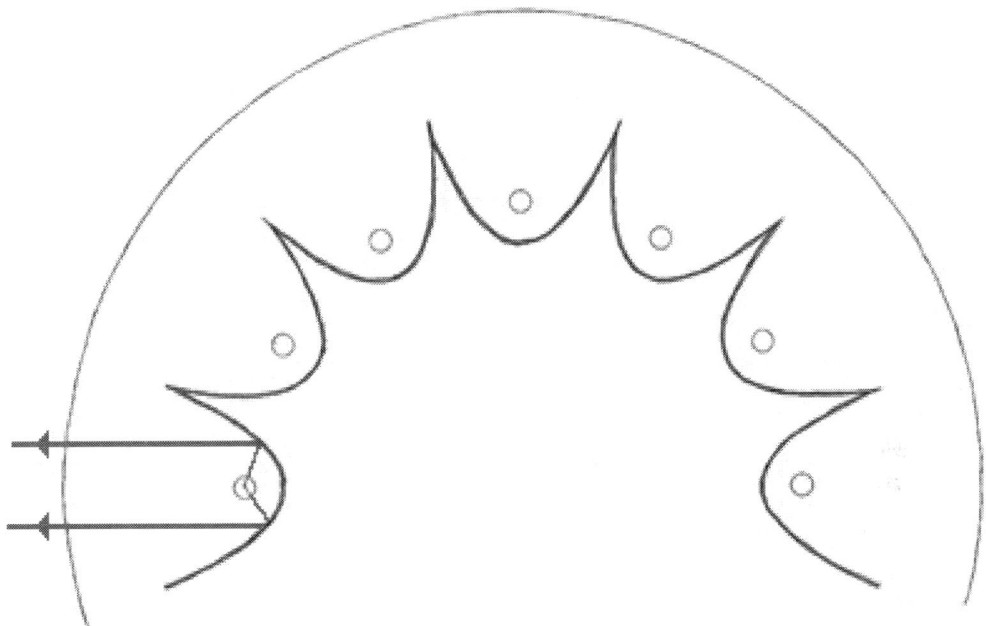

In Fig. 20 it is evident that it requires at least seven lamps to cover the 180° which would be necessary for a lighthouse located on a relatively straight section of coast. It would take a further five lamps to illuminate the whole 360° giving a total of twelve lamps. Ensuring that twelve lamps were kept supplied with fuel and their wicks were trimmed to ensure that they burned properly would keep a lightkeeper very busy. It might seem that seven lamps would be more than adequate, but the lighthouse at Sunderland, erected in 1802, required nine lamps so that it could be seen by a ship approaching the harbour from any direction.[28] Even this number is dwarfed by the thirty lamps required by the Belle Toute lighthouse which commenced operation in October 1834.[29] The system of using reflectors is known as the catoptric system and with high quality equipment maintained by skilled lightkeepers, for the first time it offered the means to create a light sufficiently powerful to give mariners adequate warning of impending danger.

While the catoptric system made a significant proportion of the light useful, it did not ensure that all the available light was making a useful contribution, too much of the light was still heading up towards the sky or down towards the sea. The answer to this problem was to use a glass lens, but before we look at the use of lenses in a lighthouse, we need to understand at least a little about the process of refraction. When a ray of light travelling in air enters a denser transparent medium there is a change in direction. The principle is shown in Fig. 21.

Fig. 21   Refraction –   Air to Glass

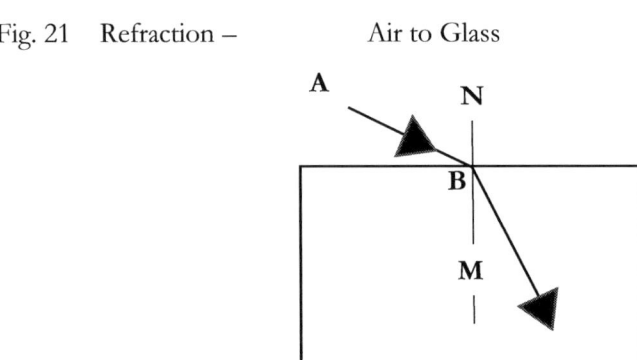

The ray AB is travelling in air and strikes the surface of the glass at B where a 'normal,' line NB has been drawn. The angle between the ray of light and the normal is called the angle of incidence **i**, just as it was when we looked at reflection. However, as glass is transparent, the light ray goes into the glass rather than being reflected. Glass may be transparent, but it is much denser than air and consequently the ray of light from B does not continue in a straight line, but moves in a direction closer to the normal NBM. The angle between the ray from B and the normal BM is called the angle of refraction **r** and you can see that angle **r** is smaller than angle **i**. The relationship between the angles of incidence and refraction is governed by Snell's Law and is called the 'refractive index.' However, we now need to look at what happens when the ray of light emerges from the glass back into the air. This is shown in Fig. 22.

Fig. 22   Refraction -   Glass to Air

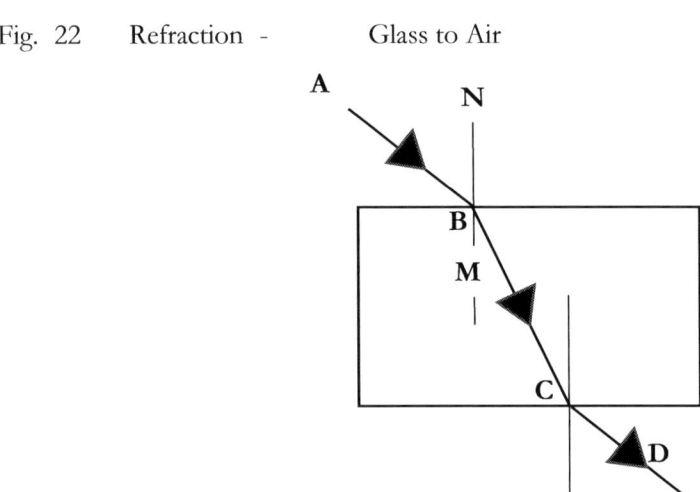

In this case, as the light is moving from the dense medium of glass to the less dense medium of air, the light is refracted in the opposite way, away from the normal and the angle of refraction **r** is larger than the angle of incidence **i**.

If we combine Figs. 21 and 22 but use a curved piece of glass, then a ray of light that would otherwise have simply gone up towards the sky is refracted twice so that it is now directed out towards the horizon where it could be seen by a mariner.

The piece of curved glass is a lens like the one in a magnifying glass and everyone is familiar with the fact that such a lens will focus the parallel rays of light from the sun to a single point. In the case of the lighthouse, it simply works the other way around.

Fig. 23   The Path of Light Rays through a Lens.

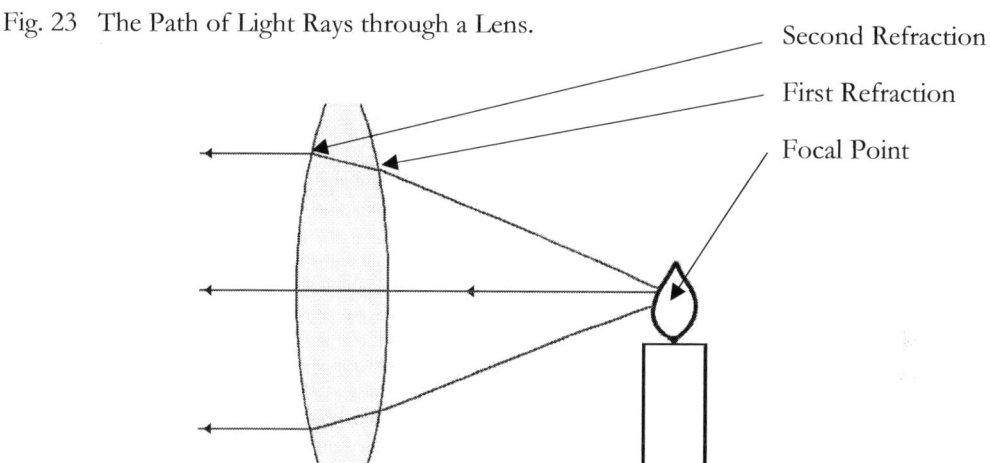

The source of light is placed at the point called the focal point of the lens and the result is a beam of light that can be directed towards the horizon. It is important to note that in this case it is the light initially going away from the source that is used to make the beam, whereas with the reflector it was the light that was originally going in the other direction. It should also be noted that it is only at the surface between the glass and the air that refraction occurs. In effect, as far as the light is concerned, the glass between the surfaces serves no useful purpose.

This arrangement would produce a strong beam of light, but there was still a problem to be overcome. If the diameter of the lens is increased, then in order to keep the surfaces at which the important refraction takes place at the correct angle, then the glass has to be made thicker in the middle. Taking an ordinary magnifying glass as an example, it is 100 mm in diameter and 15 mm thick in the middle. If it was to be made ten times greater in diameter, then it would also become ten times thicker in the middle, 150 mm thick and it would also be very heavy. Moreover, in the early nineteenth century it was very difficult to produce glass of a consistent quality in such a thickness and so the use of lenses to produce the broad and powerful beam required for a lighthouse was restricted.

A solution to this problem was developed by the French engineer Augustin-Jean Fresnel in 1821. He built on an idea propounded by Georges-Louis Leclerc, Comte de Buffon, who had suggested surrounding a central convex lens with a thin rings of glass the sides of which were ground to the same shape as they would have been on a large conventional lens.[30] Buffon's concept is shown in Fig. 24.

Fig. 24   The Comte de Buffon's Concept.

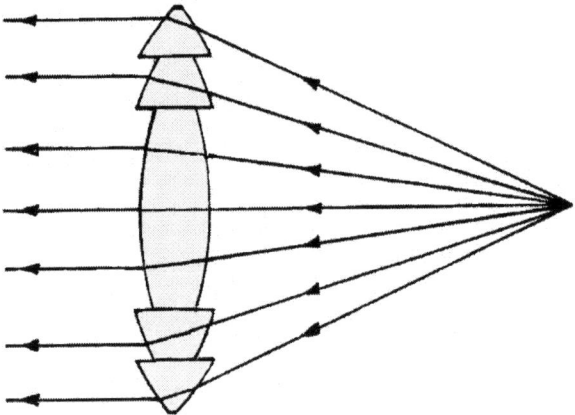

However, Buffon was only able to produce very small lenses in this way as he was trying to use a single piece of glass, whereas what was needed was a very large lens. The idea might have been valid, but the technology required to make it was lacking.

Fresnel initially tried to simplify the Comte de Buffon's idea. He retained the idea of making the lens from a single piece of glass but with one side flat so that the amount of grinding required to achieve the correct shape would be reduced, but again the glassmakers were unable to turn his idea into reality.

A much-simplified version of a Fresnel lens is shown in Fig. 25.

Fig. 25   Fresnel's Concept.

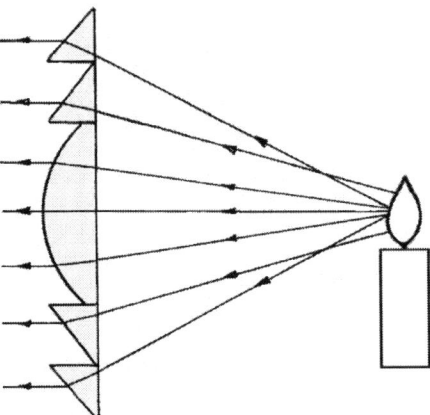

Eventually, Fresnel realised that if he was to have a lens big enough to be useful, then it would have to be made in pieces. However, while the idea might look simple on paper, the difficulty lay in making the annular prisms, the pieces of glass surrounding the central lens. These were big pieces of glass that had to be ground and polished to the exact shape required. It required much experimentation and labour before the necessary technique was finally mastered.[31]

The Fresnel lens could be made much larger by simply adding more concentric annular prisms, each with its shape adjusted slightly so that its surfaces presented the same angle to the light as the surface of a giant curved lens would have done. Fresnel's new lens system offered such obvious advantages that it was soon adopted for lighthouse illumination. It became known as the dioptric system. However, there was a limit to the size of the lens made in this way and although it produced a much more intense beam than the reflectors which had preceded it, much light was still being lost.

One solution to this was to combine the dioptric system with rings of plane mirrors, a system called catadioptric. Fig. 26 shows a catadioptric lighting apparatus producing a fixed light as installed in the lighthouse on Heugh Head, Hartlepool in 1847.[32]

Fig. 26 The Optical Apparatus from Heugh Lighthouse, Hartlepool. (Drawing by Jo Nicol).

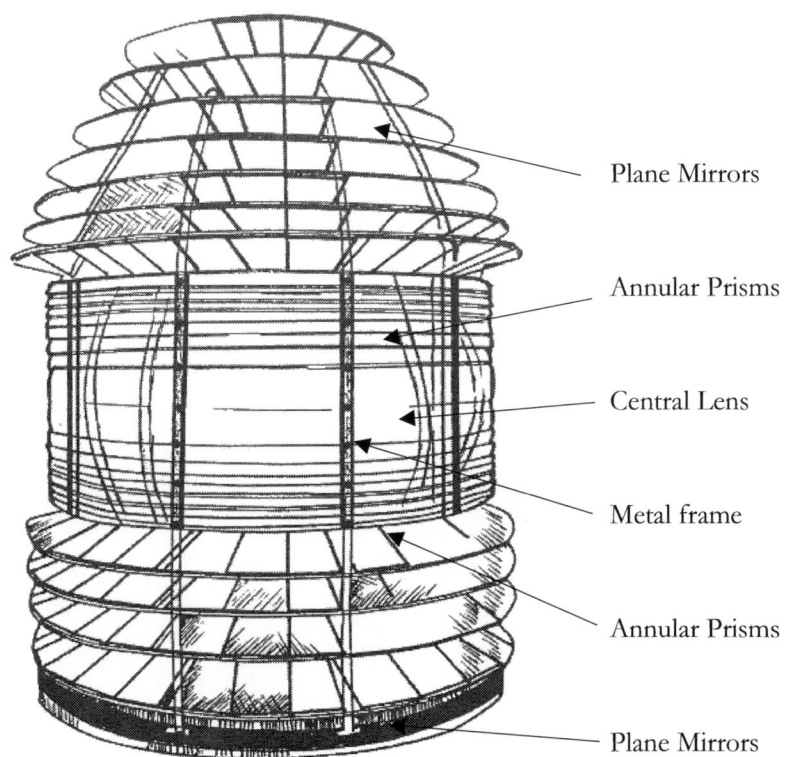

47

This optical apparatus, now preserved in the Hartlepool Museum, was used at Hartlepool's Heugh Lighthouse to provide illumination over the full 360° and so all the available light was being usefully employed, although only a very small proportion of the total light was available in any one direction. Consequently, the brightness of the light in any one direction was restricted and hence the effective range of the light was limited. Nevertheless, it provided a very effective fixed light. It should be noted that the pieces of glass are fitted into a sturdy metal frame

There was another way in which Fresnel's lens could be arranged and this is shown in Fig. 27. In this instance there is a central circular lens surrounded by annular prisms mounted in a metal frame. This arrangement would produce a beam of light in one direction.

Fig. 27   Fresnel Lens Arranged to Produce a Beam of Light.

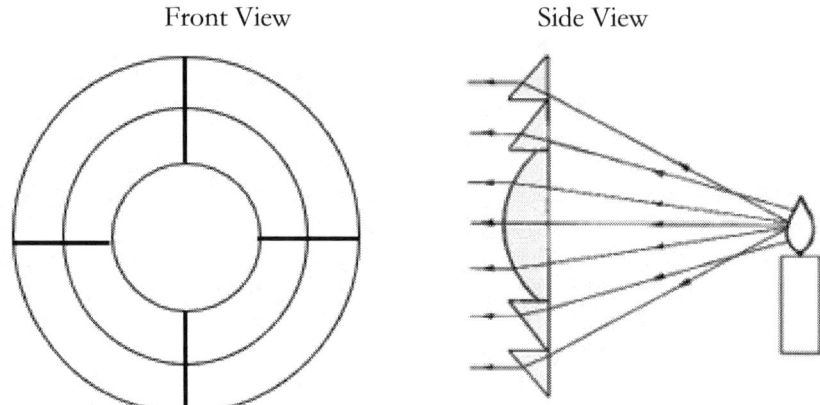

The Fresnel lens made effective use of the light already going towards the horizon, while the parabolic reflectors made use of light going away from the horizon. In each case half the light was not used. The obvious way to make the useful beam stronger would be combine the two systems in some way. However, because the lens is placed between the light source and the horizon, every ray of light must pass through the lens to reach the horizon. It is, therefore, not just a simple matter of combining the two systems. Instead of reflecting the light from the flame into a parallel beam, this time the light needs to be reflected back towards the flame and thence through the lens. The principle is illustrated in Fig. 28.

Fig. 28   Reflection plus Refraction.

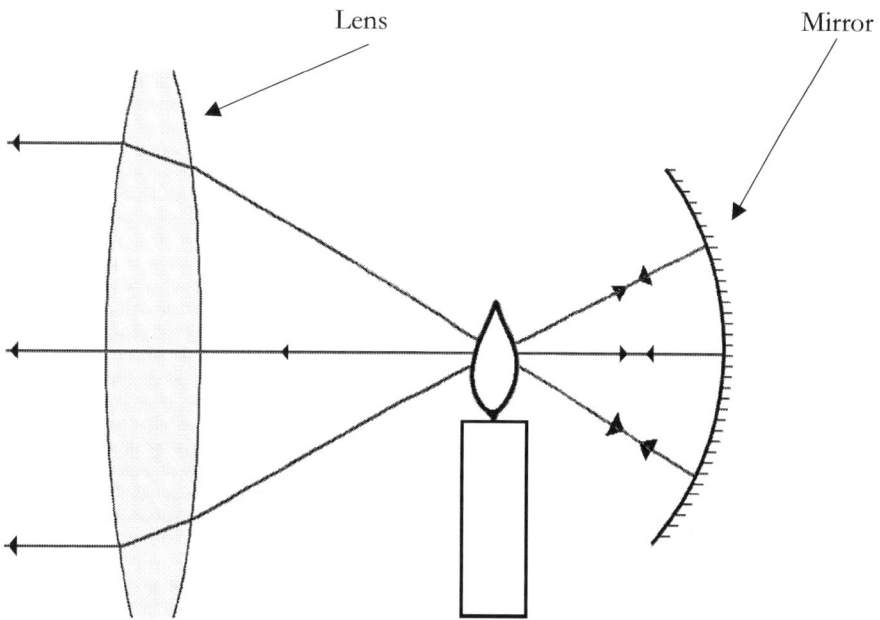

The rays of light moving to the left are refracted by the lens and form a beam that is directed towards the horizon. The rays of light that initially are moving to the right strike the mirror, which in this case is spherical in shape and has the light source at its centre. This means that every ray strikes the mirror at 90° and will, therefore, be reflected back along the same path. They will pass through the light source and then be refracted by the lens. In this way the amount of light that is directed towards the horizon is more or less doubled.

While mirrors reflected the light, no matter how carefully they were made or maintained, some of the light was lost during the reflection. It was difficult to keep the exposed polished surfaces of the mirrors clean and even the act of regular polishing could leave minute scratches that led to loss of light. A more efficient form of reflector was required.

In Fig. 22 a ray of light was shown moving from glass to air. The angle with the normal at which the ray leaves the glass is larger than the angle with the normal as it approached the glass-air boundary. Obviously, as the angle of the incident or approaching ray with the normal increases, there will come a point where the angle at which it leaves is 90° and cannot get any bigger as the ray is now parallel with the glass-air boundary. If the angle at which the ray approaches the boundary does get bigger, then the light cannot pass through the glass-air boundary and is simply reflected back as if the internal surface of the glass was a mirror. This is shown in Fig 29. The triangular shaped piece of glass within which the reflection takes place is called a prism. The prism is arranged so that the light enters the glass at right angles to the surface so it is not refracted and also leaves the glass at right angles to the surface so again there is no refraction.

Fig. 29   Total Internal Reflection.

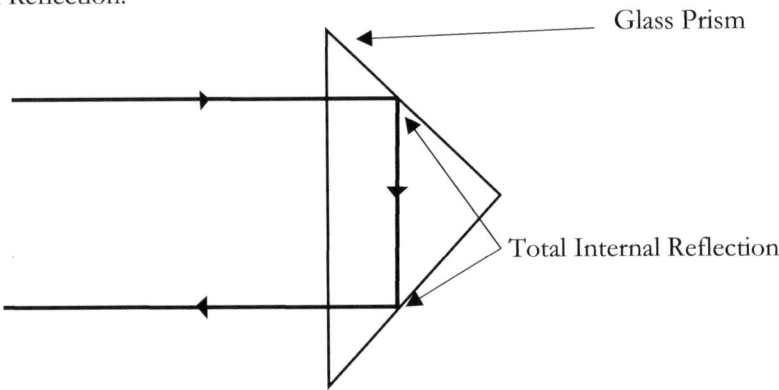

The glass-air boundary acts just like the plane mirror shown in Fig. 16. This process is called 'total internal reflection' and offered a crucial advantage over a conventional mirror. The reflecting surface is 'inside' the glass so it does not get dirty and hence needs no regular polishing, while because, unlike an ordinary mirror, almost no light is absorbed, it is much more efficient and the resultant beam is that much brighter.

The use of prisms to replace mirrors behind the light source did much more than reduce the need for cleaning and polishing. It provided a way to effectively utilize the light which would otherwise have passed around the outside of the lens system. In the catadioptric system, like that used at Hartlepool, this light had been partially utilized by a series of plane mirrors, with all their inherent defects. Prisms utilizing total internal reflection enabled Fresnel to dispense with mirrors entirely and make efficient use of this light. The result was a far more powerful beam. A Fresnel lens now consisted of a circular central section like a conventional magnifying glass, surrounded by concentric circles of prisms. Almost all the available light could now be concentrated into a uni-directional beam. It was the final step in resolving the technological problem and was called the holophotal system.

The way in which this phenomenon could be used to enhance the strength of the light beam is shown in Fig. 30. It must be stressed that this is a very much simplified drawing intended only to illustrate the principle.

Fig. 30   The Holophotal Principle.

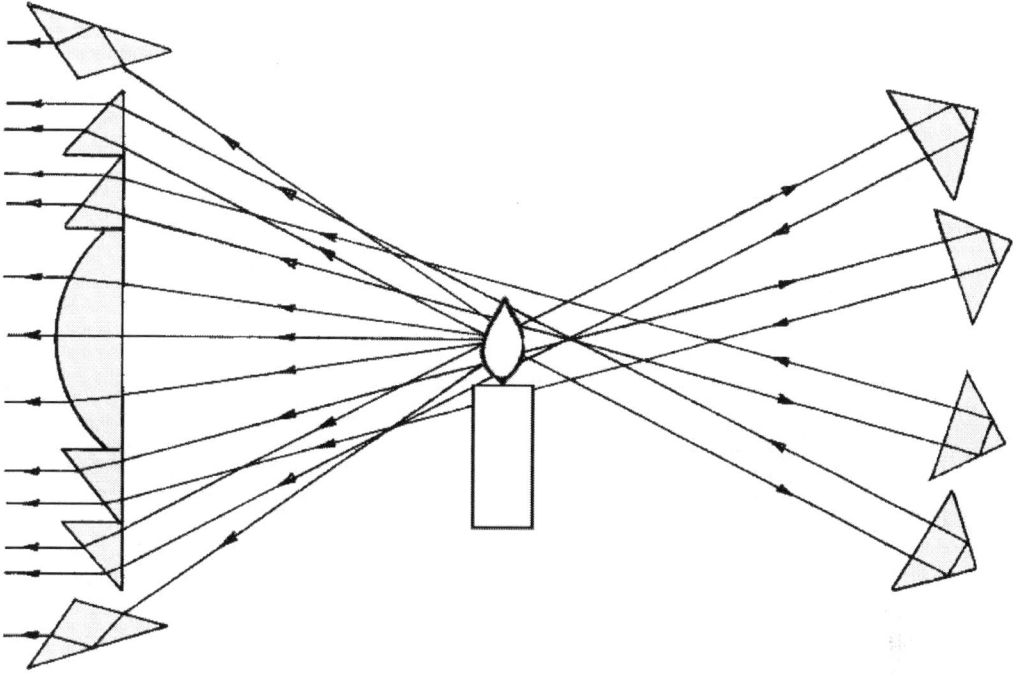

The combination of refraction and total internal reflection using just glass, not only ensured that far more of the light from the light source was effectively used, but it allowed the size of each part of the whole apparatus to be increased. The light source could now be an Argand burner with up to five concentric wicks which produced a very large amount of light, sited at the focus of an optical system that could be 6 feet (1.8 metres) tall.

The three critical elements required to produce a powerful light beam could now be combined. Argand's lamp produced an intense source of light; total internal reflection ensured that far less light was lost, while Fresnel's lens system produce a powerful uni-directional beam of light. It was at last possible to utilize enough of the light produced to create a powerful beam of light that could readily be seen over twenty nautical miles away, even in stormy conditions. The technological problems had been overcome and it was now possible to develop a comprehensive coastal illumination system.

[27] A Description and List of the Lighthouses of the World – Alexander George Findlay, Richard Holmes Laurie, London 1861.
[28] Moving Sunderland's Lighthouse 1841 – Ian Hills, FWD Publishing, 2019.
[29] The Story of the Belle Toute Lighthouse – Rob Wassell, RW Publications 2010.
[30] A Short Bright Flash – Theresa Levitt, W. W. Norton, New York, 2013.
[31] A Short Bright Flash – Theresa Levitt, W. W. Norton, New York, 2013.
[32] The Lighthouse and the Battleship – Ian Hills, FWD Publishing, 2018.

Chapter Five
What Light is That?

When there were very few lighthouses, then unless the mariner was a very considerable distance from where they thought they were, when they saw a light they could be reasonably certain which lighthouse it came from. However, as the number of lighthouses increased, so did the risk that one lighthouse could be mistaken for another with potentially catastrophic results. One common solution was to use a pair of fixed lights close to one another, while an alternative was the use of a coloured light. However, it was quickly discovered that there was a major disadvantage with the use of coloured lights, they were far less intense and hence unlikely to be seen at great distances. In fact, the only coloured light that was really effective was red. Nevertheless, no matter how it was achieved, some way to identify a specific light was essential.

There are innumerable stories about the use of false lights by wreckers who, by exhibiting lights, sought to lure vessels to their destruction, many relating to the activities of wreckers around the coasts of Cornwall. In his book Cornish Seafarers, A. K. Hamilton Jenkin[33] claims to have found only one documented example of the misuse of lights in this way and that refers to the misuse of the official Trinity House light on St Agnes in the Scilly Isles. It is alleged that they (the keepers) 'used their light to assist, rather than hinder, their relations engaged in the family occupation of wrecking.' However, while Hamilton Jenkin's assertion may be correct insofar as the absence of other substantiated evidence is concerned, it would seem to be less accurate in relation to the incident quoted above. The original source for the information was a reference in British Lighthouses, Their History and Romance by W. J. Hardy[34], a reference taken up by other historians on whose work Hamilton Jenkin relied. Hardy's assertions are analysed in detail by Cathryn Pearse[35] who demonstrates that the facts about an incident in 1681 were presented in such a way as to create a 'romantic' story and it was this which allowed speculation to creep into the narrative.

In reality, it would have been very difficult for wreckers to have used false lights to draw ships to the shore for the simple reason that they would have found it very difficult, for all the reasons noted in chapter four, to produce a light that could be seen at any great distance. Thus, the hypothesis that wreckers could exhibit a false light that would induce the captain of a vessel well out to sea to approach a dangerous shore is difficult to substantiate if only for the fact that the light from a lighthouse was largely intended to dissuade mariners from approaching the danger it marked. If there was misuse of lights, then it can only have been for vessels which were already very close to the shore and potentially already in very great danger.

The most obvious way in which a particular lighthouse could be identified simply from its light was to replace a continuous steady light with one that flashed. Then the time between flashes could be used to identify the particular lighthouse. One fairly obvious way to make the light appear to flash would be to turn it on and off at specific intervals, but as long as the light source was a flame then this was not a viable option and only became viable with the advent of electrical illumination. An alternative way to 'turn off' the light would be to use some form of mechanically operated cover that obscured and revealed the light at regular intervals. However, while this was possible, it could make it harder for the lightkeeper to access the light source.

A simpler way to make the light appear to flash was simply to make it revolve, something that became much easier with the introduction of the dioptric system. The basic principle is illustrated in Fig 31. Two lens systems are mounted in a frame which revolves around the light source. The lenses produce a concentrated beam which will appear much brighter to the mariner than the low intensity light that is otherwise visible. If the frame is made to revolve at a fixed rate by means of a clockwork apparatus, then the mariner will see a light that brightens and then dims again at regular intervals.

Fig. 31 The Revolving Light Concept.

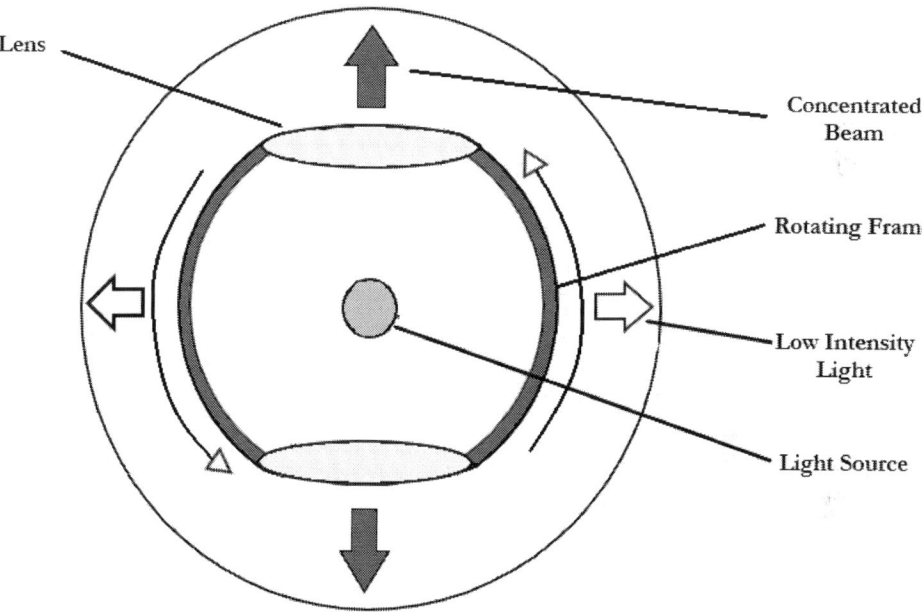

In reality, many lights were similar to the Hartlepool light illustrated in Fig. 26, but with a section of the centre panel replaced with a lens system. Either the centre section of the apparatus or the whole apparatus could be made to revolve. They would appear to the mariner as a fixed light that flashed at regular intervals. Fixed lights that incorporated a flash and revolving lights generally revolved quite slowly, perhaps once a minute, but while they provided a means to differentiate between lights, they could only utilize the light travelling in the direction of the observer and hence their intensity was limited. Nevertheless, they were employed in very large numbers.

Fig. 32   A Fixed Light with a Panel producing a Flash of Light as it Rotates.

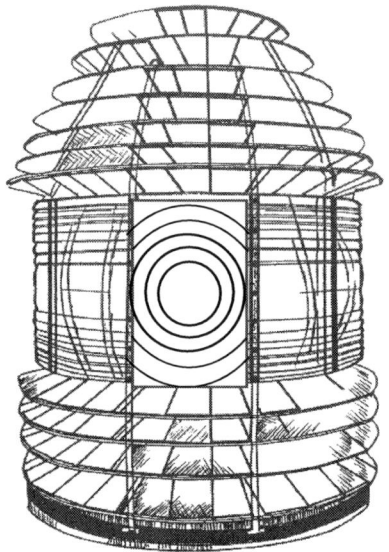

The development of the dioptric system which used total internal reflection to direct light back towards the source and then a lens system to focus it into a narrow beam, created a much more powerful beam of light in a particular direction and was called the holophotal system. It offered a way to create a very bright light in a particular direction. If the whole optical apparatus could be made to rotate, then once in every revolution there would be a very bright beam in each direction, appearing to the mariner as a flash of light. The optical apparatus would usually be designed to make multiple flashes every revolution, so by varying the number of flashes per revolution and the speed of revolution, it was possible to create a large number of different patterns. A much-simplified drawing of how a holophotal apparatus works was shown in Fig. 30.

Towards the end of the nineteenth century many of the major lighthouses were being equipped with huge, complex holophotal systems that utilized almost all the light from a large light source consisting of an Argand burner with up to four or even five concentric wicks. The huge optical apparatus which weighed many tons was often floated on mercury to create an almost frictionless bearing and rotated by means of a clockwork mechanism. The power for the clockwork rotation mechanism usually took advantage of the fact that the light was mounted at the top of a tall tower and utilized a heavy weight. Energy was stored as the weight was wound to the top of the tower and released as it descended, powering the rotational mechanism. Winding the weight to the top of the tower was just one of the duties of the lightkeeper, but a particularly onerous one.

"When the keeper came out from inside the lantern, he went to the kitchen and moved in and out from there, keeping an eye on the station, particularly on the light, and looking out to sea to check if any ship was in difficulty. Then every 30 minutes, he returned to the lantern to wind up the weight."[36]

The constructional costs of the lighthouse itself could be considerable, especially as they were often, of necessity, sited in remote locations that were difficult of access. The optical equipment itself was complex and difficult to manufacture, hence it too was expensive. In 1855 the cost of the equipment for even the least powerful apparatus, known as a Fifth Order light using silver plated reflectors and often used as a harbour light, was £226-8-0 (226 pounds 8 shillings 0 pence or £226.40)[37] equivalent to around £24,500 today. Moreover, this was the price of the equipment at the factory, it still had to be moved to the site of the lighthouse and installed. A holophotal light of the First Order could be over 8 feet (2.44 metres) high and about 6 feet (1.8 metres) wide and weigh many tons. It would, therefore, not be unreasonable to expect it to cost at least ten times as much as a Fifth Order apparatus, while twenty times as much would probably be a more realistic estimate. Unfortunately, the book, from which this information was taken[38] which is, in effect, the sales catalogue for the Chance Brothers Glass Works, does not give a price for every type of apparatus.

Fig. 33   A Holophotal Optic for a Major Lighthouse.

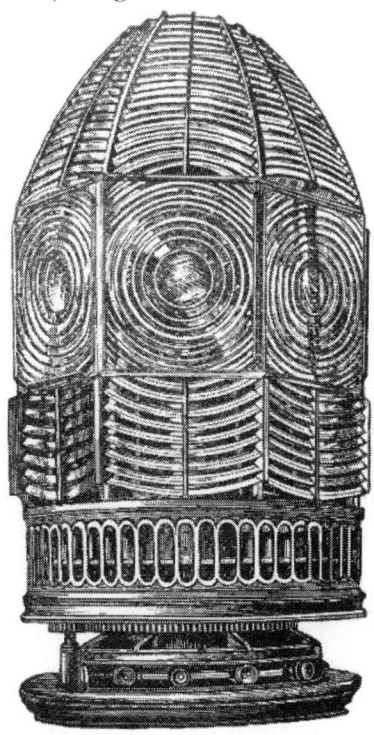

It would obviously be necessary for the mariner to know whether a particular light was fixed or revolving and if it was revolving, then how frequently and herein lies a vital role for Findlay's Lighthouse Lists. The mariner approaching land at the end of a voyage needed to be absolutely sure which light they could see, especially if they were navigating unfamiliar waters and the most effective means of doing so would be consult Findlay's List. If they were approaching the area of the coastline they expected, then they could look up the appropriate light in Findlay's List and would know its exact position, its height above the sea and how it would appear to the mariner. This would indicate whether the light was fixed or revolving and if it was a revolving light, then how frequently it revolved. In the case of a flashing light it would describe how the flashes would appear, e.g., one flash every ten seconds or two flashes every twenty seconds. This description was known as the 'characteristic' of the light.

If the light they observed matched the characteristic they were expecting, then their position was confirmed and they could take the appropriate navigational decisions. On the other hand, if the characteristic of the light they saw was unfamiliar, then they could consult Findlay's List to find out precisely which light it actually was. Once the light had been identified, then the ship's position could be determined and any necessary alteration of the ship's course calculated and implemented.

Imagine that you are in charge of the *Margaret Smith* in 1867 as she approaches the British Isles, bound for London with a cargo of sugar from Demerara. The weather has been bad for the last two weeks, so no celestial observations have been possible and it has been necessary to rely on dead reckoning, so consequently the ship's exact position is somewhat uncertain. If your dead reckoning is correct, then the ship should be headed towards the English Channel with Cornwall to the left or port side. The first light you might expect to see would be the Bishop's Rock which marks the westerly extremity of the Scilly Isles and according to Findlay's List is a fixed bright light visible at a range of 16 miles.[39] Your chart, illustrated in Fig. 34, shows the position of the lights.

As darkness falls the weather remains overcast but visibility is quite good, the sea is fairly calm and the ship is sailing before a light westerly wind. Shortly after midnight, the lookout, stationed at the masthead, reports that they can see a light ahead and just to the right or starboard of the ship, a light that gets brighter for a few seconds at two-minute intervals. It cannot be the Bishop's Rock light because that is a fixed light, whereas the characteristic of the light reported is that of a revolving light. Clearly the position you calculated by dead reckoning is wrong. You had made an allowance in your calculations for a northerly drift but if that allowance had been too big then you could be much further south and the light is that on the island of Ushant off the coast of France.

Fig. 34 Dead Reckoning.

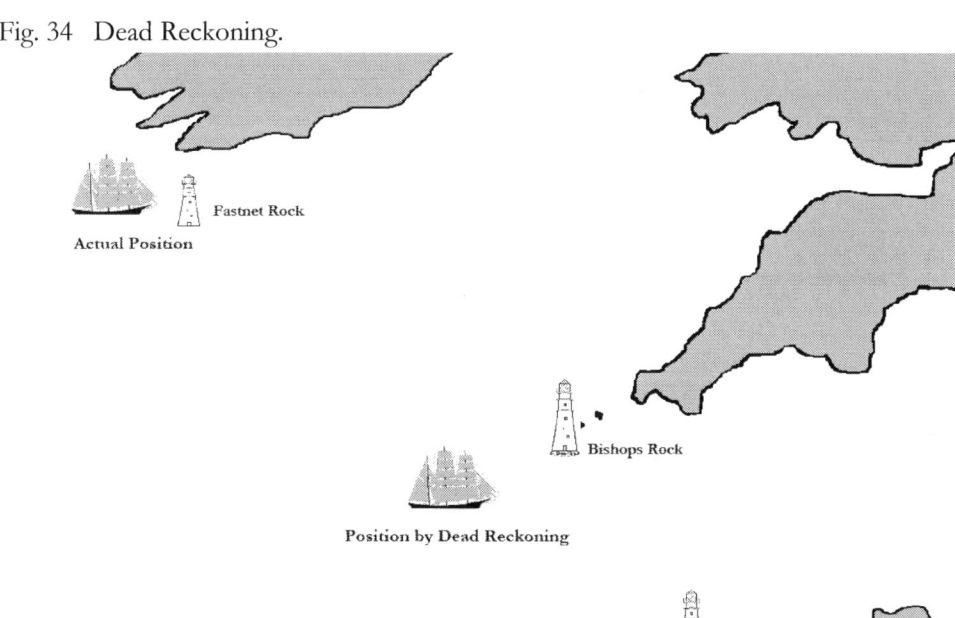

However, according to Findlay's List the Ushant lighthouse exhibits two lights, one being a fixed light and the other a revolving light which gets brighter at intervals of 30 seconds, visible at 24 miles.[40] This does not fit with the characteristic of the light reported by the lookout, so perhaps your allowance for the ship's northerly drift was too small and you are actually much further north. If that was the case, then you might expect to see the light from Fastnet Rock, which, according to Findlay's List is a revolving light with a brighter flash every two minutes, visible at 18 miles.[41] That fits with what the lookout reports, but if Fastnet is to your right or starboard, you are heading directly for the coast of southern Ireland. Fortunately, in this case the weather was kind, visibility was good, so Fastnet Rock would be some 18 miles ahead and the lookout's prompt warning would enable you to make the necessary alteration of course with little difficulty. However, if the conditions had been stormy with poor visibility it could have easily become a very dangerous situation.

In this example the *Margaret Smith* was around 185 nautical miles further north and about 200 nautical miles west of your position as calculated by dead reckoning. It would seem to be a very big error, but if you had been forced to rely on dead reckoning for 14 days, which is 336 hours and had underestimated the ship's northward drift by an average of only 0.6 of a knot, then that would be more than enough to account for the error in latitude. Similarly, if you had overestimated the ship's speed, again by an average of just 0.6 of a knot, then that would easily account for the error in longitude.

[33] Cornish Seafarers - A. K. Hamilton Jenkin, 1932.
[34] British Lighthouses, Their History and Romance – W. J. Hardy, Religious Tract Society, London, 1895.
[35] Neglectful or Worse – Cathryn Pearse in Troze, the Official Journal of the National Maritime Museum, Cornwall, 2008.
[36] The Lightkeeper – Gerald Butler, Liffey Press, Dublin, 2012.
[37] Description and Plans of Lights for Lighthouses – Chance Bothers – Birmingham 1855.
[38] Description and Plans of Lights for Lighthouses – Chance Bothers – Birmingham 1855.
[39] A Description and List of the Lighthouses of the World, Seventh Edition 1867 – Alexander George Findlay, Richard Holmes Laurie, London, 1867.
[40] A Description and List of the Lighthouses of the World, Seventh Edition 1867 – Alexander George Findlay, Richard Holmes Laurie, London, 1867.
[41] A Description and List of the Lighthouses of the World, Seventh Edition 1867 – Alexander George Findlay, Richard Holmes Laurie, London, 1867.

Chapter Six
The Beginnings of Findlay's Lighthouse Lists

Alexander George Findlay was a renowned hydrographer[42] who, by the middle of the nineteenth century, had established a reputation for producing valuable aids to navigators and had demonstrated a profound interest in lighthouses. His work would demonstrate the growth in the number of lights established to guide the mariner and also mark the technological developments that made lights so much more effective.

At the end of the eighteenth century there were just 37 major lighthouses around the coasts of the British Isles, almost all of them displaying a fixed bright light.[43] This number had more than doubled to 82 by 1832, but the percentage of fixed lights had fallen to around 63%. The number would double again by 1861 although the percentage of fixed lights would remain almost constant, it was 62% in 1861[44], showing that many of the new lights were adopting some of the techniques described earlier to ensure that they could be differentiated from others close by. These figures clearly demonstrate that there had been a major advance in the lighting of the coasts around the British Isles over the first half of the nineteenth century.

However, as much use will be made of numerical data from Findlay's Lists and other sources, it is important to understand the basis for these numbers. They are all derived from either Admiralty Lists or Findlay's Lists and refer only to major lights. In all these publications, a distinction was drawn between major lights and minor ones. In general, major lights were those which were visible at a range of 10 nautical miles or more and were the first lights that a mariner would see as they approached the coast. In addition to these major lights there were numerous other minor lights which would enable the navigator to make their way into a harbour, through a narrow channel or into an anchorage. While undoubtedly important for the navigator, they have not been included in the data used as their sheer number would obscure many important points. In every case the data used is based on what either the Admiralty or Findlay deemed to be a major light.

Prior to the 1832 Admiralty List and Findlay's List there had already been an attempt to provide mariners with a list of all the major lights around the British Isles, the title page of which is shown in Fig. 35.[45]

Fig. 35   Frontispiece of The British Pharos.

THE BRITISH PHAROS;

OR A LIST OF THE LIGHTHOUSES

ON THE COASTS OF GREAT BRITAIN AND IRELAND,

DESCRIPTIVE OF THE

APPEARANCE OF THE LIGHTS AT NIGHT.

FOR THE USE OF MARINERS.

SECOND EDITION.

LEITH:
PUBLISHED AND SOLD BY W. REID & SON;
AND TO BE HAD OF ALL THE CHART-SELLERS
IN THE UNITED KINGDOM.

MDCCCXXXI.

It was attributed to Alan Stevenson of the famous family of lighthouse engineers and the second edition was published by W. Reid & Son, Leith in 1831. It is worth noting that it states the book is available at all the chart-sellers in the United Kingdom. The format is wholly different in that it eschews a tabular arrangement in favour of a prose description. An example of the style is shown below.

TYNEMOUTH CASTLE LIGHT

Situate on the northern side of the entrance to the River Tyne, in the county of Northumberland.

HERE the light revolves, and is seen at the distance of five or six leagues*, and at less distances in an obscure state of the atmosphere. The light appears, in its brightest state, once in every minute, like a star of the first magnitude; but gradually becoming less luminous, it is at length eclipsed.

*A league was three nautical miles or 5.56 km.

It should be evident that this verbal description contains only a limited amount of information, a location based on a description, the fact that the light revolves once a minute and the distance at which it could be seen. Nevertheless, it must have been of some value to mariners as it ran to two editions.

The Admiralty maintained a hydrographic office which produced all the materials that a navigator would need. These would include charts, light lists, pilot guides and the mass of numerical tables associated with solar or celestial navigation. It is a role that continued until the end of the twentieth century when the organization became known as the United Kingdom Hydrographic Office[46] and it continues to this day.

The Admiralty list for 1832[47] recognized the fact that most of the lights were fixed and that in order to distinguish them there were variations in colour or the actual number of lights, thereby giving prominence to these features. Altogether, there were up to fifteen separate data entries for each light. The format of an entry is illustrated in Table 1. As fifteen columns required a very wide page, entries were actual spread across two facing pages. In Table 1a the first eight columns are shown, the remaining seven are illustrated in Table 1b. In addition to the fifteen data columns, each light was given a sequential number (omitted in Table 1) starting with 1 for the Isles of Scilly, then working anti-clockwise around the British mainland.

Fig. 36   Frontispiece of the 1832 Admiralty List.

THE

LIGHT-HOUSES

OF

THE BRITISH ISLANDS,

CORRECTED TO APRIL,

1832.

———

HYDROGRAPHICAL OFFICE, ADMIRALTY.

LONDON:
PRINTED BY WILLIAM CLOWES, 14, CHARING-CROSS.

MDCCCXXXII.

Table 1a

Name of Light	Point or Place where on the Light-house stands	Number of Lights, with their Magnetic Bearing and Distance from each other	Fixed, flashing or revolving	Time of Revolution or Flash	Colour of the Lights	Distance in Miles at which they are easily seen in clear Weather
LIZARD	Lizard Point	Two N.N.W. ½W. 223 feet	Fixed	……….	Bright	20

Table 1b

Points of the Magnetic Compass between which they are shown	If Harbour Lights, the time during which it is lighted	Colour of the Light-house, or any peculiarity in its appearance by Day	Height of Lantern above the Sea at High Water in Feet	Height of Building from Lantern to Base in Feet	Year erected	Remarks
E.N.E. ¼ E. seaward to N.W. by W. ½W.	…………	White	East 221 W. 224	Both 45.	1751	When in one, lead clear of the Manacles and Shags.

It is interesting to note that the Admiralty list does not give the latitude and longitude of the light, relying instead upon a verbal description. This description was, presumably, directly related to the nomenclature used on the relevant Admiralty chart. In other words, to make use of the Admiralty Lights list it might well be necessary to also have the corresponding Admiralty chart and it is unlikely that these were a cheap item. There was thus scope for other organizations to produce charts of their own which could be offered commercially, effectively in competition with the Admiralty. The situation is the same today where Admiralty charts and those produced by other commercial publishers are readily available. A few moments research will demonstrate that Admiralty charts tend to be more expensive.

It was noted at the start of the chapter that in the decade prior to the publication of Findlay's first list of Lighthouses of the World in 1861, there had been a rapid growth in the number of lights. This would require mariners to constantly update the information they held and there can be little doubt that to have all the information to hand in a single volume that encompassed the whole world would have been very helpful. There was, therefore, a commercial opportunity for a publisher who could offer a comprehensive list of lights that was not explicitly linked to any specific set of charts. Findlay's publisher, Richard Holmes Laurie, also recognized that mariners would appreciate having the most up-to-date information possible. They realized that this would inevitably be a quite expensive publication but could, in part, justify the cost by offering free updates, obtained by sending a coupon to the publishers.

The first edition of Findlay's list of Lighthouses of the World contains an introduction to what he calls pharology, the description of lighthouses and their illumination. It gives a brief history of lighthouses and how they are constructed, together with a discussion about how they are operated. It also emphasises the fact that the hazards in the waters around the British Isles are, to a much greater extent than any other country except the United States, marked by a very considerable number of lightships, despite the fact that these vessels are expensive to maintain.

In fact, much of the introduction is comprised of extracts from two papers delivered by Findlay to the Society of Arts in 1847 and 1858. Particular attention is given to explaining the different types of lighthouse illumination, their characteristics and the respective advantages and disadvantages. This introduction, ostensibly to provide the mariner with information as to the methods by which the lights are produced, would also have been a potential selling point for the non-mariner who had an interest in the subject.

Fig. 37   Cover from 1867 Edition and Frontispiece from 1861 Edition.

The actual information about the lights in Findlay's Lists was provided in a common format throughout the nineteenth century and is shown in Table 2.

Name and Character of the Light	Latitude N Longitude W	Description etc.	Description of Apparatus	Height above High water	Visible in Miles	Year established
Table 2						
FLATHOLM One bright fixed light	51 22.5 3 7.0	A white tower, 80 feet high, on the S point	1a	38	17	1839
Major lights were printed in block capitals to distinguish them from minor Lights. The entry below is for the lighthouse on the island of Flatholm in the Bristol Channel and is taken from the 1861 edition of Findlay's List.		This described what the mariner would see in daylight.	The number indicated the size or order of the light and the letter the type of optical equipment used.			

The content of the columns is readily apparent except for column 4 which provides a code for the nature of the apparatus. The key to the symbols as provided by Findlay is given in Table 2a.

Table 2a	
*	signifies a catoptric, or reflector light.
1a, 2, 3d, etc.	indicate dioptric, or lens lights, the figure showing the order or size, 1st, 2nd, 3rd, to 6th order*
a	a fixed lenticular** light.
b	a revolving lenticular light.
c	a fixed and flashing light.
d	a holophotal light.
Note	* The largest and most powerful apparatus was a 1st order light and a 6th the smallest. ** Lenticular means constructed with lenses, to create a magnifying effect.

The information gave the mariner the location of the light, its appearance by day, the characteristic of its light by night and the range at which it could be seen, all of which would enable it to be recognized. However, there is one situation where a complication arises. The key determinant of the distance at which a light could be visible is the height of the light above the sea, but this assumes that visibility conditions are perfect. When atmospheric conditions were less than perfect, the distance at which the light could be seen was heavily influenced by the power of the light source and the size of the lenticular apparatus. Thus, while a small, fifth order light could theoretically be visible at twenty miles, poor visibility conditions could easily reduce that to five miles; whereas a larger, much more powerful, first order light might easily be visible at fifteen miles. Similarly, a type **d**, holophotal apparatus, would be expected to produce a much more powerful and penetrating beam than a type **c**, fixed and flashing apparatus.

This information was, of course, of crucial importance to the mariner approaching the coast. It was of vital importance for their safety that they had available accurate and up-to-date information, but here there was a problem, for the information was, as we have seen, constantly changing. We live in a time when the dissemination of information can be done on a global basis in a fraction of a second. In the nineteenth century things moved much more slowly. Until the development of the electric telegraph, information travelled as a 'physical object', either written on a piece of paper, or carried in the head of a human. Inevitably, this meant there was a considerable time lag between an event and news of that event reaching a wide audience. In the case of information about lighthouses, the information had to first be transmitted to a central point from which it could then be disseminated.

In areas with access to good communications the information could be received at the central point fairly quickly, certainly in a few days. It would also be possible to provide advance warning of changes, e.g., a new lighthouse was to be built at this place and was expected to operate from a certain date, exhibiting a light which flashed every 15 seconds. However, every country operated independently and while some shared information very effectively, this was not necessarily true of all. However, no matter how efficient the organization, once the information had to cross the seas then the transmission times would be very significantly extended. It should be remembered that in those days an Atlantic crossing in less than a month was considered to be fast while it could take two to three months to sail from Britain to places such as India or Australia.

Another issue that needs to be considered is the source of information. In Britain it was fairly straightforward. Trinity House could supply the information for England and Wales, the Northern Lighthouse Board covered Scotland while the Commissioners of the Irish Lights dealt with Ireland. All of them fed data to the Admiralty where it could be marked on charts and incorporated into the Admiralty List of Lights. Indeed, Findlay makes it clear in the quotation shown below that he makes extensive use of the Admiralty's information.

The list of Lights which follow have been re-arranged from those published by the Admiralty, which, under the careful superintendence of Commander Edward Dunsterville, R.N., have obtained a completeness approaching perfection.[48]

Given that all this information was already available, then it might be reasonable to ask why Findlay's Lists were ever published. They were not published by some philanthropic organization, but by a commercial publisher, Richard Holmes Laurie, who it can safely be assumed would not have undertaken such a project unless they could discern a profitable market for it. Moreover, their product was by no means a cheap one. The price, three shillings and sixpence, is clearly marked on the front of the 1867 edition. Converting three shillings and sixpence to modern decimal currency gives 17.5p which is the equivalent of almost £20 in today's money, although it must be noted that at this time any bound book was an expensive item.[49]

---

[42] Hydrography is the study of the physical features of the earth's navigable waters and the coasts that border them.

[43] The Light-Houses of The British Isles, Hydrographical Office, Admiralty, London 1832

[44] A Description and List of the Lighthouses of the World – Alexander George Findlay; Richard Holmes Laurie, London, 1861.

[45] The British Pharos – Alan Stevenson; W. Reid & Son, Leith, 1831.

[46] The United Kingdom Hydrographic Office is part of the Admiralty and is based in Taunton, Somerset. www.gov.uk/government/organizations/uk-hydrographic-office

[47] The Light-Houses of The British Isles, Hydrographical Office, Admiralty, London 1832.

[48] A Description and List of the Lighthouses of the World – Alexander George Findlay; Richard Holmes Laurie, London, 1861.

[49] Bank of England Calculator. https://www.bankofengland.co.uk/monetary-policy/inflation/inflation-calculator

Chapter Seven
The Coasts are Lit

Findlay published his first list of Lighthouses of the World in 1861. This reflected the fact that there had been a substantial rise in the number of major lighthouses around the British Isles in the first half of the nineteenth century. A similar situation applied to the coasts of Europe and the countries surrounding the Mediterranean, although in this case much of the development had occurred in the preceding decade. This is illustrated in Table 3 below. Throughout the rest of the century Findlay's lists chart the continued and substantial growth in the number of lights on every coastline of the world.

Table 3	Increase in the Number of Major Lights between 1851 and 1861		
Geographical Area	No. of Lights	No. built or modified in last 10 years	Percentage of new or modified Lights
	Findlay's Lists, like all similar lists, identify major lighthouses as distinct from minor lights such as harbour lights. Some lightvessels were also classified as major lights.	This includes completely new lighthouses and lightvessels, together with older structures where the characteristic of the light as seen by the mariner had significantly changed. Thus, for example, this could include the replacement of a previously fixed light with a revolving one.	Percentages are approximate as a small proportion of lights in the list have no date so while they are included in the overall number, they may be omitted from the number of new and modified lights. The percentage may therefore be lower.
British Isles	112	11	10
Europe - White Sea to Gibraltar	210	63	30
Mediterranean Region	122	66	54

If we made the assumption that a similar rate of change could be applied to all the minor lights, an assumption that almost certainly incorporates a significant underestimation, as there were far more minor lights than major ones, the number of changes would have been very much higher. Thus, Findlay's List offered mariners a very real advantage for no matter where their voyage might take them, they could find all the changes in a single, readily portable volume.

It is clear from Table 3 that the coasts of Europe and at least the northern shores of the Mediterranean had a significant number of major lighthouses. This is hardly surprising given the fact that sea trade routes had existed here for hundreds of years and that in many European countries industrialization, with all its impact on the volume of trade, was well advanced. However, the situation in other parts of the world was rather different.

Table 4	Increase in the Number of Major Lights between 1851 and 1861		
Geographical Area	No. of Lights	No. built or modified in last 10 years	Percentage of new or modified Lights
Africa, Atlantic Islands, Red Sea and Indian Ocean	30	10	33
Far East inc. East Indies, China, Philippines & Pacific	7	2	29
Australasia	23	13	56
Eastern North America	162	119	73
Caribbean	18	7	39
South America	22	11	50
Western North America	14	13	93

It is evident from Table 4 that the development of lighthouses across the rest of the world was at a far earlier stage of development, the one exception being the eastern seaboard of North America. Marine traffic between Europe and North America was extensive for it must be remembered that this was a period when large numbers of Europeans were emigrating in search of a better life. Additionally, the eastern seaboard of the United States made extensive use of the oceans for freight traffic as well as a major fishery industry. Indeed, especially in the north-eastern states, there was a large and flourishing maritime industry. However, the coast of New England in particular was rugged with numerous small islands and shoals so the need for an effective lighthouse system had been there for some time.

Lighthouse development in the rest of the world is shown as being at a much earlier stage of development, but the figures must be treated with a degree of caution. There would inevitably be a very considerable delay between the establishment of a light in a far-off place such as the East Indies and the information about that light reaching London. Thus, the information would always tend to be a little out-of-date and it is perfectly possible that some lights existed that were not included in Findlay's list. However, it is also clear from the data in Table 4 that significant development was underway. The full extent of that development is demonstrated in Table 5. The geographical areas used in this table and subsequently, are derived from Findlay's lists where, with only a few exceptions, lights were listed in this order.

In order to keep the number of areas manageable and to facilitate analysis of the data, some areas, have been coalesced into larger groupings.

Table 5	Major Lights – Development to 1899				
Geographical Area	No. of Lights				
	1861	1867	1879	1890	1899
British Isles	112	143	158	191	206
Europe - White Sea to Gibraltar	210	224	260	312	349
Mediterranean Region	122	160	196	257	291
Africa, Atlantic Islands, Red Sea, Indian Ocean	30	53	74	93	115
Far East inc. East Indies, China, Philippines & Pacific	7	14	54	96	139
Australasia	23	32	52	86	98
Eastern North America	162	179	192	225	235
Caribbean*	18	22	35	46	54
South America	22	23	43	62	71
Western North America	14	15	20	24	38
Note	* Includes Bermuda which Findlay does not include with the Caribbean but with the Atlantic Islands in the Africa, Red Sea and Indian Ocean area.				

However, while there was an increase in the number of major lights in each geographical area, there are also very wide variations. This becomes even more obvious if we look at the comparative growth. If we compare the area with one of the largest number of major lights in 1861, the British Isles, with the area that had the lowest number of major lights, the Far East, then the difference in the growth rate becomes obvious.

Fig. 38   Growth Comparison between British Isles and Far East.

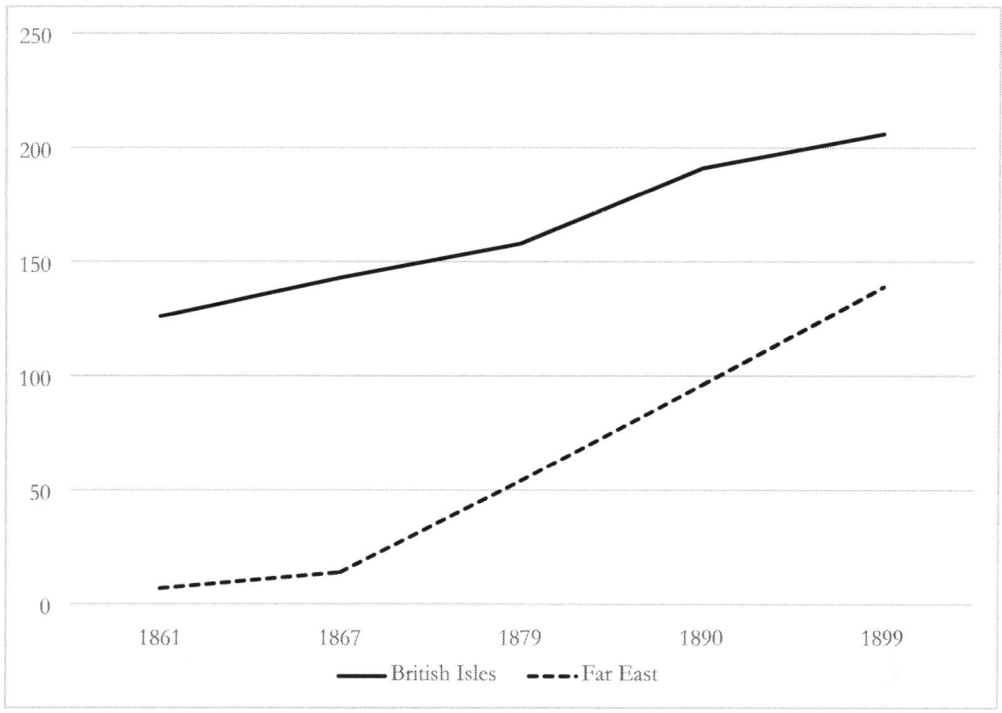

The rate of growth in the Far East and Pacific has clearly been far greater than that for the British Isles but despite this, the overall number of lights is still lower. If the length of coastline is taken into account, then the coasts of the Far East might still be considered as under-lit. However, it is also important to take into account the fact that while the seas around the British Isles were almost all part of well-frequented shipping lanes. In the Pacific area there were very busy shipping lanes around China and Japan, but also vast areas where there was little maritime traffic.

The demand for lights was clearly influenced by a whole range of factors. Consider the different growth rates for Africa, Red Sea and Indian Ocean compared with the Mediterranean.

Fig. 39   Growth Comparison between Mediterranean and Africa, Red Sea & Indian Ocean.

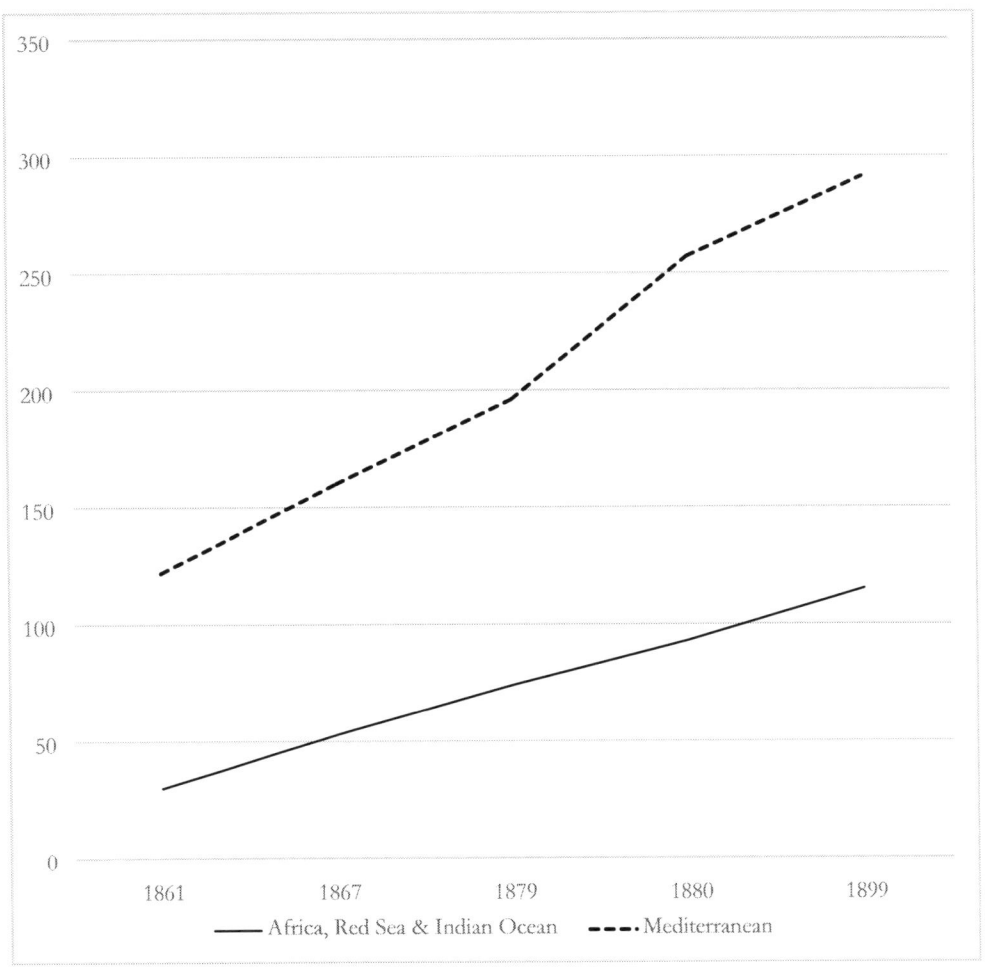

The data for the Mediterranean shows a rate of growth that is faster than the growth for Africa, the Red Sea and Indian Ocean; while there is a growing disparity in the overall number of lights, which, given the very significant difference in the length of coastline would, at first sight, seem surprising. However, as the demand for lights was invariably linked with the density of trade traffic, there is one very obvious reason for the discrepancy. The opening of the Suex Canal in 1869 ensured that an increasing proportion of the vessels sailing to and from India, the Far East and Australasia, especially the steamships which now carried the prestige traffic, took this route rather than the longer passage round Africa. This stimulated the provision of navigational aids throughout the Mediterranean. It also had an impact on lights along the Red Sea coasts, but the distances involved and hence the number of lights required was smaller. The route around the Cape of Good Hope was frequented now generally by sailing vessels for which the Suez Canal was not a realistic option. Moreover, at this time most of Africa was colonial territory controlled by European powers and thus Britain controlled a substantial length of the coastline. Colonial administrations were always unwilling to incur expenditure which did not result in a direct benefit to themselves and consequently had little interest in providing navigational aids for ships which were simply sailing past.[50]

Another way to compare the growth rates for major lights in different parts of the world is to look at the overall percentage increase over the 48 years from 1861 to 1899. This is shown in Table 6.

Table 6	Percentage Growth of Major Lights (Highest – Lowest) 1861 – 1899.		
Geographical Area	No. of Lights in 1861	No. of Lights in 1899	Percentage of additional Lights
Far East inc. East Indies, China, Philippines & Pacific	7	139	1,886 %
Australasia	23	98	326 %
Africa, Atlantic Islands, Red Sea and Indian Ocean	30	115	283 %
South America	22	71	223 %
Caribbean	18	54	200 %
Western North America	14	38	171 %
Mediterranean	122	291	139 %
Europe	210	349	66 %
British Isles	126	206	63 %
Eastern North America	162	235	45 %

As a key element in determining the percentage growth is the baseline or number of lights in 1861, it is unsurprising that those areas which had fewer lights in 1861 exhibit the higher percentage growth rates. The coasts of Europe, the British Isles and the eastern seaboard of North America were already quite well illuminated by 1861 so there was less need for such a large increase in the number of lights. Nevertheless, the increase in these areas was far from insignificant.

The relationship between the number of lights in 1861 and the growth by 1899 may be more easily recognized if the information from Table 6 is presented in graphical form.

Fig. 40

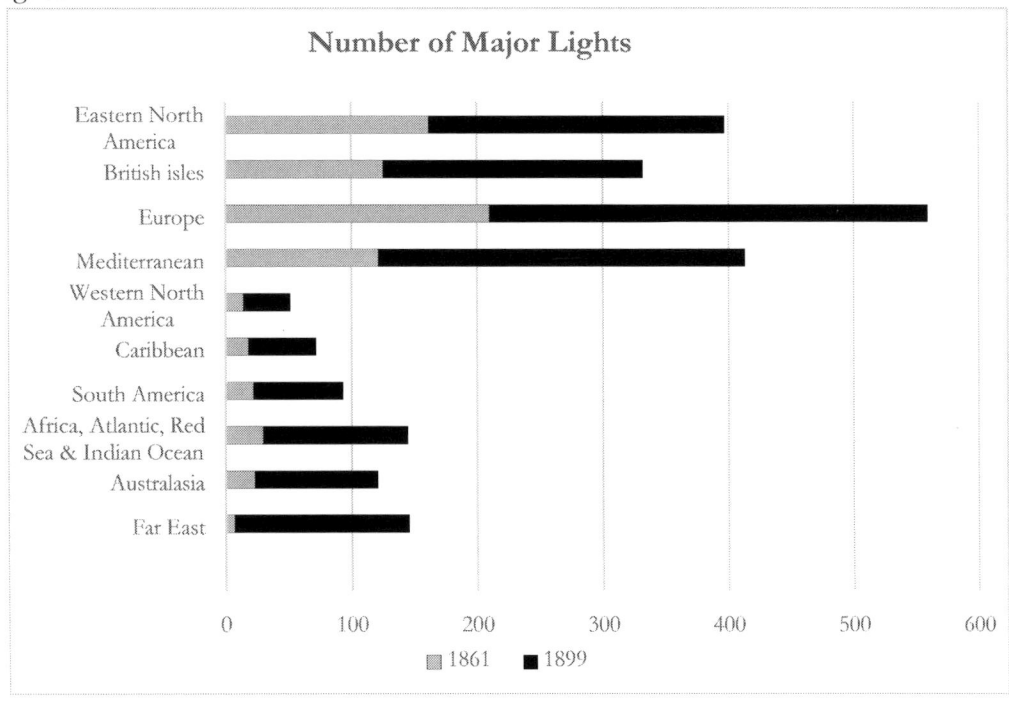

Fig. 41

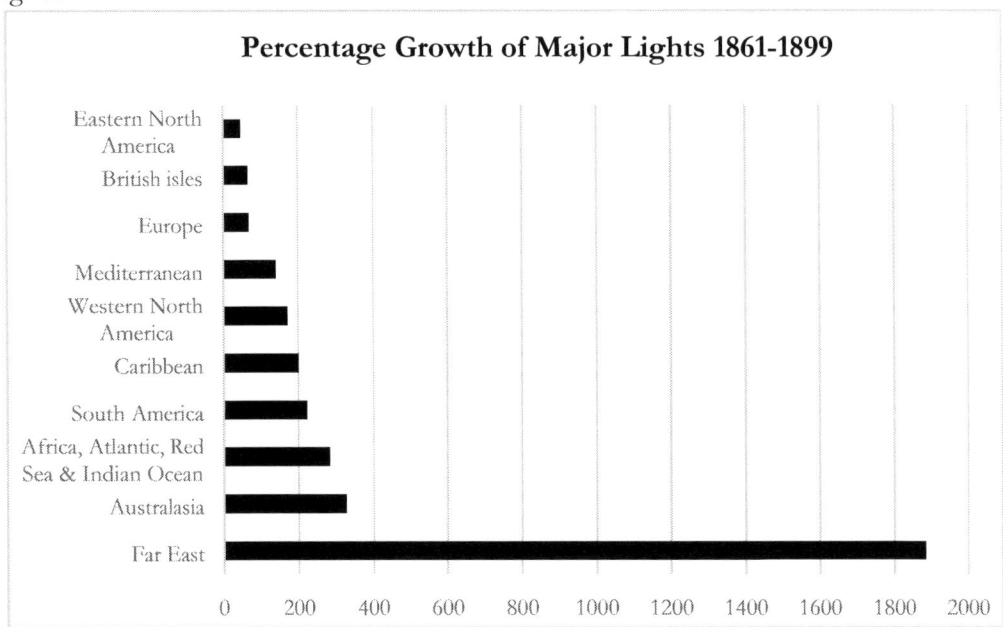

Hitherto, all the data used has related to the 'major lights' as defined in the Admiralty and Findlay's lists, with no distinction being drawn between land based and floating lights. Table 7 shows the number of light vessels classed as major lights in each geographical area over the period 1861 to 1899.

Table 7 Light Vessels classified as Major Lights					
Geographical Area	No. of Light Vessels				
	1861	1867	1879	1890	1899
British Isles	14	14	27	42	42
Europe - White Sea to Gibraltar	2	5	13	20	21
Mediterranean Region	0	0	1	1	0
Africa, Atlantic Islands, Red Sea, Indian Ocean	0	2	0	0	0
Far East inc., East Indies, China, Philippines & Pacific	0	1	2	0	0
Australasia	0	1	1	2	0
Eastern North America	3	3	4	6	10
Caribbean	0	0	1	1	0
South America	3	3	1	1	3
Western North America	0	1	1	0	2

The data in Table 7 clearly illustrates that only the British Isles, Europe and the eastern seaboard of North America made extensive use of Light Vessels. A reasonable question is thus, can a reason for this disparity be deduced? To answer the question it is first necessary to consider where a light vessel would be needed, to which the answer must be that it is required to mark some underwater hazard such as a sandbank, or to indicate a deep-water channel. In either case it is obvious that a light vessel is associated with shallow water. However, two further obvious criteria for the establishment of a light vessel in a particular location would be the volume of traffic and the risk posed by the underwater hazard. The necessity of marking a deep-water channel implies that there are areas of shallow water close by.

This situation is very well illustrated by the problems for mariners created by the Goodwin Sands, which are a major hazard in the English Channel, then, as now, one of the world's busiest shipping routes. The Goodwin Sands lie just off the south east coast of England and many hundreds of ships, nobody really knows how many, have been wrecked on them. They lie about three to five miles offshore so that lighthouses built on the land would not be very effective in warning mariners of the danger. Therefore, the only answer was to use a light vessel, effectively a ship on which a kind of mini-lighthouse was mounted. However, there was a problem because if the ship had masts to carry sails, then these would obscure the light and make the light vessel useless. Therefore, a light vessel usually had to be towed to the right place and then held there with anchors. In the case of the Goodwin Sands, even in 1861, three light vessels were stationed there to warn mariners of the danger. This is illustrated in Fig. 42.

Fig. 42   Light Vessels Marking the Goodwin Sands.

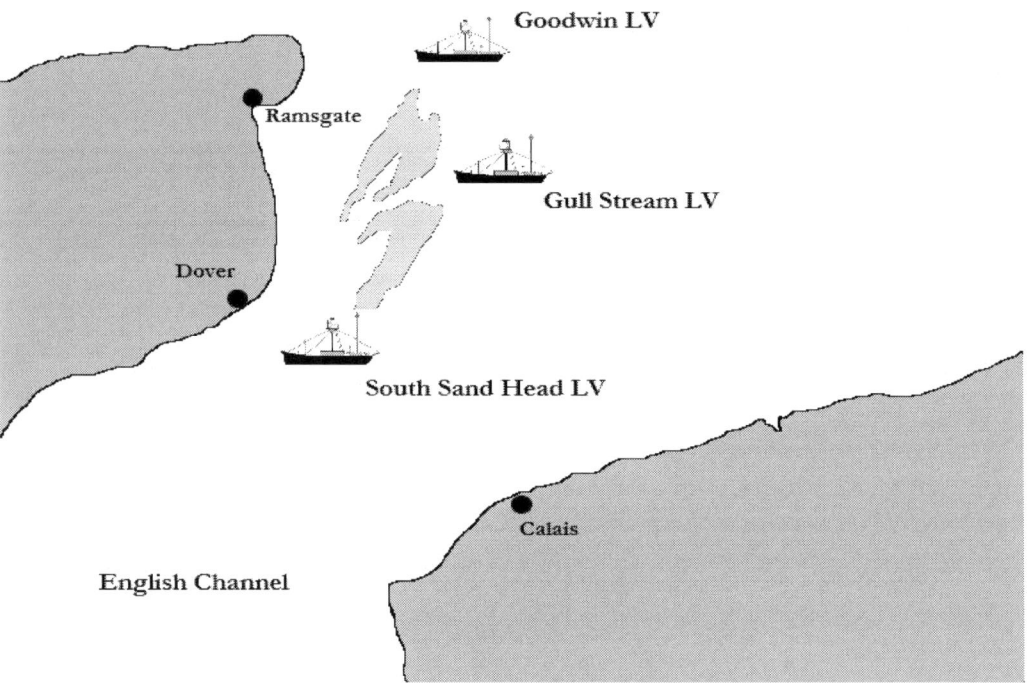

The importance attached to marking shallow waters, sandbanks or other underwater hazards may be judged from the data in Table 7 which shows that by the close of the nineteenth century there were 42 lightvessels categorized as major lights. However, Findlay's lists also show that there were many other light vessels, which while undoubtedly serving a very important purpose, were considered to be of somewhat lesser importance.

[50] Lighthouses-The Race to Illuminate the World – Toby Chance and Peter Williams, New Holland, London, 2008.

Chapter Eight
The March of Technology

Findlay's first list of lights was published in 1861 by which time the technological problems associated with the creation of a powerful and recognizable light had, to a large extent been resolved. Thus, when describing each light Findlay had to be prepared to recognize many different forms of apparatus. However, when the Admiralty List of Lights around the British Isles was published in 1832 it was evident from the format that there was an expectation that a large proportion of the lights would exhibit a fixed light, for provision was included to show the number of lights and their colour. In fact, of the 80 major lights listed, 51 of them were fixed lights of one type or another.

Table 8	Types of Optical Apparatus in Major Lights 1832[51]
Type of Apparatus	**Number**
Single Fixed Bright/White Lights	34
Two or more Fixed Lights	17
Revolving Lights	21
Flashing Lights	3
Lights which exhibited a combination, e.g., one fixed light and one revolving light	5

However, as Table 8 shows, a significant proportion of the lights were already equipped with apparatus that made it easier to create a light with a more recognizable characteristic. Given the importance of these major lights and the necessity to be able to recognize them at long range, it might be expected that the proportion of lights equipped with the more sophisticated apparatus required to produce a distinct characteristic would increase. The figures in the 1861 column of Table 9b draw upon Findlay's 1861 list and demonstrate that this is exactly what happened.

Findlay's early lists place the characteristics of each light into one of six groups. These groups, together with a definition for each group, are shown in Table 9a.

Table 9a	
Characteristic	Description
Fixed or Steady	Self-explanatory
Revolving	Light gradually increasing to full effect, and gradually decreasing to eclipse, at equal intervals of one, two or three minutes, but occasionally as often as three times in a minute
Flashing	Showing five or more flashes and eclipses alternately in a minute
Fixed light with Red or White Flash	Flash preceded and followed by a short eclipse at intervals of two, three or four minutes
Intermittent	Suddenly appearing in view, remaining visible for a certain time, and then as suddenly eclipsed for a shorter time
Red and White	Red and white light alternately at equal intervals, without any intervening eclipse

To simplify analysis, henceforward lights exhibiting any of the three latter characteristics have been considered together and described as 'Combination.'

Table 9b	Types of Optical Apparatus in Major Lights 1832 and 1861	
Type of Apparatus	No. in 1832	No. in 1861
Fixed Lights	51	78
Revolving Lights	21	30
Flashing Lights	3	14
Lights which exhibited a combination, e.g., one fixed light and one revolving light	5	4

It can be seen from the figures in Table 9b there has been an increase in almost all types of apparatus, but the type showing the highest proportional increase was the 'flashing' apparatus which, as we saw in an earlier chapter, was also the equipment capable of producing the beam of light with the greatest intensity. However, it is also evident that many additional major lights were still being equipped with apparatus to produce a fixed bright light, although the number equipped with revolving lights was also increasing. Why was this the case? If the apparatus that produced flashing lights also created more intense beams of light, then it would surely have made sense to incorporate it into every new light and to replace older equipment in every existing light.

The data in Table 10 shows the type of apparatus used in major lights around the British Isles through the second half of the nineteenth century.

Year	Fixed		Flashing		Revolving		Combination		Total
	No.	%	No.	%	No.	%	No.	%	No.
1861	78	61.9	14	11.1	30	23.8	4	3.2	126
1867	83	58.0	13	9.1	38	26.6	9	6.3	143
1879	76	48.1	15	9.5	54	34.2	13	8.2	158
1890	53	27.7	45	23.6	57	29.8	36	18.8	191
1899	47	22.8	63	30.6	56	27.2	40	19.4	206

Table 10 — Types of Optical Apparatus used in Major Lights around the British Isles

As might be expected, after a small amount of growth between 1861 and 1867, the number of fixed lights declined steadily. Moreover, even allowing for the small initial growth in numbers, the percentage of fixed lights fell continuously. In 1861 almost 2 out of every three lights was fixed, whilst by 1899 it had fallen to about 1 in five. Over the same period the total number of major lights had grown by 80 (63%) so clearly very few new lights could have been equipped with 'fixed' apparatus. However, surprisingly, it shows that the number of flashing lights, those using the most effective technology, did not really begin to grow substantially until the last twenty years of the century.

There were clear advantages offered by the latest technology, which was certainly available throughout the period. The French had equipped the Cordouan tower with such apparatus in 1823 where it replaced twelve large reflectors. In the next fifteen years the French had equipped twenty lights with this modern apparatus and were clearly way ahead of the British in the application of this technology.[52] Given the importance of lights to Britain, this low rate of growth requires some explanation.

Even the most cursory glance at the holophotal apparatus shown in Fig. 30 should be enough to convince anyone that this was a very complex piece of equipment. It consisted of hundreds of pieces of glass, each of which had to be accurately ground and polished to achieve a very precise shape. Unsurprisingly, there were very few glassmakers with the ability to do this and they had to employ some very highly skilled workers as well as specialized equipment. Building the optical apparatus for lights was a highly specialized sector of the glass industry and required investment and resources on a significant scale. No company would be prepared to invest the necessary capital unless they could be reasonably certain of achieving an adequate return and this required having the requisite expertise and a ready market. It should, therefore, come as no surprise that many of the installations for lights were built in France. After all, the man who originally developed the idea, Augustin Fresnel, was French and there was already an established optical glass industry. Moreover, France had a long and, in many cases, dangerous coastline to protect and so there was the potential of a ready market to hand.

Meanwhile, in Britain which, as we have seen, had by 1832 already established a significant number of lighthouses, the manufacture of suitable lenses was in its infancy. The

Northern Lighthouse Board had imported a lens from France and had also requested a British company, Messrs. Cookson & Co., of Newcastle, to produce a lens. In 1833 they conducted a trial of the French lens, the Cookson lens and their own reflectors. The results were conclusive. The lenses produced light equivalent to seven reflectors.[53]

The British lighthouse authorities were anxious not to be dependent upon any foreign supplier, especially a French one, so in 1834 they asked the Cookson Glass Company to begin Fresnel lens production. Cookson's promptly brought in Léonor Fresnel, Augustin's brother, as a consultant and by 1836 they had overtaken the French in this technology. Cookson's made a few lenses, but they were unable to surmount some of the very real difficulties encountered in making this type of equipment. After experimenting with moulding and grinding, in 1845 they eventually sold the business, leaving the French with a virtual monopoly.[54]

However, another British company, one that was to be most closely associated with this optical equipment for lights, Chance Brothers of Smethwick, took up the challenge. In 1832 Chance Brothers had asked M. Georges Bontemps, Director of Choisy-le-Roi glass, to advise them on sheet glass manufacture, taking advantage of the French lead in this technology. However, James Timmins Chance rapidly developed a mechanized grinding machine that reduced the manufacturing time and cost of sheet glass. The company specialized in flat, sheet glass for windows and in 1848 had successfully won the contract to supply almost a million square feet of glass for the Crystal Palace which was to house the Great Exhibition.

Building on the successful arrangement with Georges Bontemps, Chance Brothers persuaded him to resign from Choisy-le Roi and move to Smethwick in order to assist them in setting up the process needed to make optical glass of the appropriate quality for dioptric elements and Fresnel lenses. They could then begin to compete with the already established French companies. This should have meant that around the British Isles the number of flashing lights utilizing the latest technology rose sharply, but as we can see from Table 9b that did not happen. Clearly there was a problem.

In fact, there were several problems. Some of these were associated with difficulties in the manufacturing process. Each element of the apparatus had to be made to a very exact specification which made the grinding and polishing process very slow and expensive. Furthermore, a bureaucratic concern to avoid any one company gaining a monopoly meant that Chance Brothers were not even allowed to tender for the large and complex gunmetal framework required to mount the optical elements, although they had, of necessity, to build one in order to construct and test the apparatus. The issue was expressed forthrightly in a book published by Chance Brothers about their lighthouse work and their frustration is evident.

> They were not even allowed to tender for the metal frame-work, although obliged to construct such for their own use in adjusting the glass. And it is

clear that under this system, when the lenses and prisms came to be set up at the lighthouse in a different framework and by other hands, the pains taken by the constructor to ensure accuracy of adjustment might be thrown away. Or, as happened in some cases, an unsuitable lamp might be provided, or the bars of the lantern be placed so as to intercept some of the light. Nor could allowance be made for the 'dip' of the horizon.[55]

In fact, there is one small part of the quotation above that points us to an issue that seriously retarded the development of the holophotal system. Whilst, in theory, it should have produced a far more intense beam, there were complaints that it was, in reality, less effective than the old reflector lights. When the complaints were investigated, it was found that they were, to a large extent, justified. There was clearly a mismatch between theory and practice and this would have to be investigated.

James Chance was heavily involved in this investigation which took place between 1859 and 1861. The importance attached to it may be deduced from the fact that it also engaged the talents of eminent men such as the Astronomer Royal, Sir George Biddel Airy and Professor Faraday. After much experimentation it was found that the effectiveness of the new optical apparatus was very closely dependent upon the accurate location of the light source, even a difference of a quarter of an inch (6.4 mm) made a huge difference. The most critical adjustment was the height of the flame in relation to the optical apparatus. If the flame was too low, then most of the light was directed to a point well above the horizon and virtually useless to mariners; too high and light was directed into the sea well before the horizon. The accuracy of the adjustment required to align all the parts correctly had hitherto simply not been appreciated. No longer could the light be considered as a kind of flat-pack parts kit that could be assembled in a rough and ready way by whoever was available, but as a complex whole which needed to be installed and adjusted with meticulous attention to detail. In his paper 'On Optical Apparatus used in Lighthouses,' presented to the Institution of Civil Engineers in May 1867, James Chance outlined the test and adjustment procedure now carried out in the factory to ensure that the light was as efficient as possible.[56] However, in the same paper he also castigates the Lighthouse Authorities for their procedures used hitherto.

> Nothing could be more unscientific than the system which was, until a recent date, frequently practised by the lighthouse authorities of this country: the manufacturer of lighthouse apparatus often supplied only the separate panels with the glass permanently fixed in them; and an intervening contractor was employed to frame them together.

This revealing statement goes a long way to explain what the data in Findlay's Lists demonstrated. The British lighthouse authorities had been very slow to adjust their practices and had only begun to do so shortly before 1867. Hence one reason for the lack of progress in installing modern flashing lights between 1861 and 1867 can be presented. What is harder to explain is the continued lack of progress between 1867 and 1879. The most likely answer is simple inertia. A great deal of effort had been put into developing the best possible reflector lights. These were built by skilled and experienced craftsmen, resulting in lights that worked quite well, could be constructed for a modest sum, were reliable, simple to operate and maintain. Another issue that would undoubtedly have acted as a brake on progress was that so many lights around the British Isles were operated by a variety of different organizations. The report of the Select Committee on Lighthouses in 1845 noted that:

> The Lights have hitherto been divided into two classes, viz. the Public General Lights, which are of use to all vessels passing the coasts; and the Harbour or Local Lights, which are specifically for the use of vessels resorting to particular ports.
> The first class is now under the Trinity House in England, the Northern Commissioners in Scotland, and the Ballast Board in Ireland.

Another factor which undoubtedly had an influence was cost. Many of the older lighthouses had been built not by any public authority, but as commercial ventures where the objective was to make a profit. The proprietors of these lights would undoubtedly have been inclined to limit expenditure to the minimum and had little motive to improve their lights. The problems this caused had been recognized and under an Act of 1822, the purchase of such private lights by the established lighthouse authorities was authorized. The purchase was far from cheap. As an example, the light on Flatholm in the Bristol Channel was purchased in 1823 for the sum of £15,828-19s-4d, the equivalent of at least £1,898,000 in 2019, while in 1825 the purchase of the Fern light cost £36,000, which is equivalent to £3,389,000. Moreover, these were just the cheap lights. The cost of purchasing the Skerries light in 1834 was £444,084-11s-3d, equivalent to a staggering £58,154,586 today. Altogether, the cost of bringing the privately owned lights into public ownership was £1,182,545, which would be equivalent to £154,859,025, a big sum even today.[57]

However, despite such a big expenditure, some lights classified by Findlay as major lights remained in private hands. One example of this is the Lynas light on the coast of North Wales which was operated by the Liverpool Dock Trustees. In 1832 it had exhibited a fixed light, but by 1845 it had been converted to a flashing light. The testimony of Lt. William Lord, RN., Marine Superintendent for the Liverpool Dock Trustees provides another indication of why the more general application of flashing lights was tardy.

> 4332   Chairman] Have you, in your observations, paid any attention to the difference between flashing and fixed lights, as to their utility on the coast?
> -- I have; one of our lights is a flashing light at Point Lynas.

4333   Which do you prefer as a distinguishing light; in what instance would you employ flashing instead of fixed lights? -- To effect a distinction is a very important object, and I think a flashing light is seen at a greater distance; it has a peculiar physiological effect on the retina of the eye.

4334   Then when flashing lights can be used to advantage to discriminate from light to light, you think their effect would be good upon the shipping? -- I think flashing lights are good, but it would not do to adopt them entirely, as the only mode to distinguish them would then be by the interval elapsing between each flash, which would be very dangerous to a vessel making the light in thick tempestuous weather.[58]

This testimony seems to indicate a degree of confusion in the mind of the witness. On the one hand they say that there is an advantage in range, but then seem to argue that this advantage becomes disadvantageous in stormy weather. Clearly, they were not sufficiently convinced of the advantages of flashing lights as other major lights under their control such as Bidston still retained its fixed light in 1899.

However, no matter what the opinion of men such as William Lord, the study of optical apparatus was swiftly becoming more and more scientific. As early as 1849 Alan Stevenson had set out a series of scientific methods, backed by all the necessary mathematical calculations, to correctly align the optical system of a lighthouse.[59] In his 1867 paper to the Institution of Civil Engineers, Chance had shown the advantage of designing the optical apparatus to meet the specific requirements of a particular location. Meanwhile, just as the science was moving on, so was the technology that supported the manufacturing process. Chance Brothers developed machines to do the grinding and polishing and bought details of the processes, tools, moulding equipment and formulas required for optical quality glass from Edmond Feil who had a small optical glass factory in France. They were then in a position to produce optical equipment as good as any in the world.

We can look at the data in Findlay's Lists to see the effect of technological development in other parts of the world. Table 11 shows the type and percentage of each type of optical apparatus used in major lights in each geographic area.

Table 11		Types of Optical Apparatus used in Major Lights by Geographic Area*								
Area	Year	Fixed		Flashing		Revolving		Combination		Total
		No.	%	No.	%	No.	%	No.	%	No.
Africa, Atlantic, Red Sea & Indian Ocean	1861	22	73.3%	0	0	7	23.3%	1	3.2%	30
	1867	31	58.5%	1	1.9%	18	34.0%	3	5.7%	53
	1879	45	60.8%	3	4.1%	20	27.0%	6	8.1%	74
	1890	48	51.6%	10	10.8%	27	29.0%	8	8.6%	93
	1899	47	40.9%	24	20.9%	31	27.0%	13	11.3%	115
Australasia	1861	9	39.1%	0	0%	12	52.2%	2	8.7%	23
	1867	16	50.0%	0	0%	14	43.8%	2	6.3%	32
	1879	25	48.1%	1	1.9%	23	44.2%	3	5.8%	35
	1890	37	43.0%	10	11.6%	32	37.2%	7	8.1%	86
	1899	39	39.8%	17	17.3%	35	35.7%	7	7.1%	98
Caribbean	1861	5	27.8%	0	0%	9	50.0%	4	22.2%	18
	1867	7	31.8%	0	0%	10	45.5%	5	22.7%	22
	1879	15	42.9%	0	0%	16	45.7%	4	11.4%	35
	1890	14	30.4%	1	2.2%	21	45.7%	10	21.7%	46
	1899	16	29.6%	3	5.6%	24	44.4%	11	20.4%	54
Eastern North America	1861	98	60.5%	2	1.2%	37	22.8%	25	15.4%	162
	1867	108	60.3%	3	1.7%	43	24.0%	25	14.0%	179
	1879	100	52.1%	7	3.6%	59	30.7%	26	13.5%	192
	1890	109	48.4%	19	8.4%	63	28.0%	34	15.1%	225
	1899	100	42.6%	29	12.3%	66	28.1%	40	17.0%	235
Europe	1861	114	54.3%	0	0%	54	25.7%	42	20.0%	210
	1867	120	53.6%	0	0%	55	24.6%	49	21.9%	224
	1879	131	50.4%	13	5.0%	62	23.8%	54	20.8%	260
	1890	127	40.7%	42	13.5%	64	20.5%	79	25.3%	312
	1899	114	32.7%	82	23.5%	60	17.2%	93	26.6%	349
Far East	1861	5	71.45	0	0%	2	28.6%	0	0%	7
	1867	11	78.6%	0	0%	3	21.4%	0	0%	14
	1879	38	70.4%	2	3.7%	12	22.2%	2	3.7%	54
	1890	54	56.3%	9	9.4%	23	24.0%	10	10.4%	96
	1899	56	40.3%	30	21.6%	41	29.5%	12	8.6%	139
Mediterranean	1861	51	41.8%	1	0.8%	43	35.2%	27	22.1%	122
	1867	61	38.1%	1	0.6%	58	36.3%	40	25.0	160
	1879	75	38.3%	6	3.1%	65	33.2%	50	25.5%	196
	1890	82	31.9%	14	5.4%	65	25.3%	96	37.4%	257
	1899	78	26.8%	33	11.3%	73	25.1%	107	36.8%	291
South America	1861	9	40.9%	0	0%	10	45.5%	3	13.6%	22
	1867	8	34.8%	0	0%	11	47.8%	4	17.4%	23

	1879	16	37.2%	0	0%	18	41.9%	9	20.9%	**43**
	1890	28	45.2%	2	3.2%	20	32.3%	12	19.4%	**62**
	1899	23	32.4%	2	2.8%	28	39.4%	18	25.4%	**71**
Western North America	1861	7	50.0%	1	7.1%	3	21.4%	3	21.4%	**14**
	1867	7	46.7%	1	6.7%	3	20.0%	4	26.7%	**15**
	1879	9	45.05	3	15.05	4	20.0%	4	20.0%	**20**
	1890	9	37.5%	5	20.8%	6	25.0%	4	16.7%	**24**
	1899	12	31.6%	9	23.7%	10	26.3%	7	18.4%	**38**
Note	* Data drawn from Findlay's Lists for 1861, 1867, 1879, 1890 and 1899.									

It can be seen from the data in Table 11 that in almost every area there is a pattern of growth in the number and percentage of flashing lights and a decline in the percentage of fixed lights. This is what might have been expected on the basis that there was a greater likelihood that newly erected lights would receive the latest equipment, while existing lights, especially those with revolving lights that were already distinguishable, continued to use their existing equipment until such time as it required replacement. It is also evident that in some areas the increase in the number of flashing lights was much higher in the last decade of the century. Again, this is hardly surprising given that by 1890 Chance Brothers in England and firms like Henry-Lepaute and Louis Sautter (later Sautter Harle) had fully mastered the necessary techniques and were able to design and build the complex apparatus required with facility.

Fig. 43 Percentage of Major Lights with Flashing Apparatus.

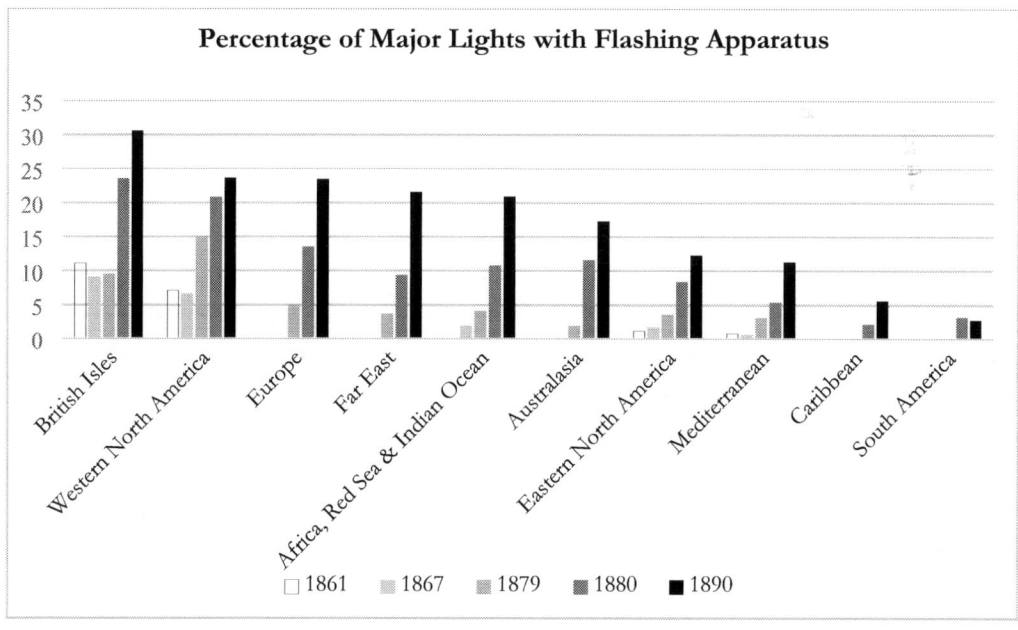

Table 11 shows that, as might have been expected, the higher percentages of major lights using the most sophisticated equipment were those located in areas where industrialization was likely to be most advanced, although on this basis the position of both Eastern and Western North America might appear to be anomalous. However, one important reason for this is that the United States had by 1859 equipped most of its lighthouses with Fresnel lenses. These were still operating satisfactorily; they gave four times the light of a reflector lamp and used only a quarter of the amount of oil.[60]

There is, however, another anomaly hidden within the data. It was noted earlier that the French had over twenty lights using Fresnel lenses by 1843, so why does the percentage for Europe in 1861 show up as zero? Did Findlay get it wrong? The answer lies in Fig. 31, which shows a revolving light system whereby a fixed light effectively had flashes superimposed upon it. As the French were the first to make Fresnel lenses, they incorporated them into their lights before the complete holophotal system was fully developed. They are described in Findlay's Lists as 'Fixed with flash' to differentiate them from a holophotal light which produced only flashes.

Another reason why the number of flashing lights does not show the increase that might have been expected, comes from the fact that for many years it was considered that flashing lights could not be employed on light vessels. The light vessel will roll and pitch when the sea is rough and if the light apparatus was rigidly attached to the ship, then instead of pointing to the horizon, one side of the light would be directed to the sky and the other to the sea. To overcome this, the light apparatus, usually a reflecting light, was suspended from gimbals so that it always stayed vertical. It was held that the very much heavier holophotal apparatus would be prone to damage and would require a stabilization system.[61] It is certainly correct that many light vessels exhibited either a single or perhaps multiple fixed lights. However, the simpler system had the inherent defect that it required multiple burners and consequently more maintenance, especially if multiple lights were shown as each would require a whole set of lamps to cover all 360° of the light vessel's horizon. However, the advantages of the holophotal system, a single lamp, a more intense beam and a readily recognizable characteristic were advantages that could not be ignored and by the end of the nineteenth century light vessels were beginning to acquire flashing characteristics.

However, there is another problem when marking the position of things like sandbanks. The sea is constantly moving the sand and hence the position of the sandbank is not fixed and can change. The advantage of a light vessel is that it is fairly easy to move if the position of the sandbank changes. Table 12 shows the position of the three light vessels warning of the Goodwin Sands in 1861 and then in 1880.

Table 12	Position of Light Vessels marking the Goodwin Sands			
Light Vessel	1861		1880	
	Latitude	Longitude	Latitude	Longitude
Goodwin LV	52 19	1 35	51 19.5	1 35.3
Gull Stream LV*	51 17	1 30	51 16	1 28.4
South Sand Head LV	51 10	1 28	51 9	1 28
Note	* In 1880 the Gull Stream LV was no longer classified as a major light. An additional lightship called the East Goodwin now marked the east of the sand and was positioned at Lat. 51 13 Long. 1 36.4.			

Table 12 shows that the position of each light vessel was a little different around 20 years later. The changes may appear quite small, but were nevertheless important as in the narrow waters of the English Channel an error of as little as a mile could have serious consequences.

Light vessels were, as Findlay points out in the 1865 edition of his lists, very expensive to operate.

> Three men are sufficient to a rock lighthouse, 11 are required to man a Lightship; consequently, while the annual cost of a first-class Lighthouse is from £265 - £340, that of the Lightships amounts to £1,103.[62]

Findlay also makes it very clear that light vessels were considered of the utmost importance to mariners, although they were fully aware of the danger to which the light vessel crew were exposed. In a storm they were already dangerously close to an underwater hazard and totally dependent upon their anchors. Findlay pays the men who manned them the following tribute.

> The Lightvessels very seldom go adrift, and there is no instance on record in which the crew have voluntarily run from their stations in bad weather. Where they have been driven from their moorings, the vessels have always been replaced in a very short time and none have ever been wrecked.[63]

Unfortunately, on the 27th November 1954 a violent storm, one of the worst storms in the English Channel for 200 years, severed the South Goodwin Lightship's mooring cable in the middle of the night. It was soon noticed that the lightship was not in her usual position and lifeboats from Ramsgate and Dover were launched. She was eventually found over six miles away, lying on her side on the sands, with waves breaking over her. A helicopter flying over the wreck saw a man, clad in pyjamas, clinging to the wreck and he was rescued. It transpired that he was actually from the Ministry of Agriculture and Fisheries rather than a member of the seven-man crew, all of whom sadly perished.[64]

[51] The Light-Houses of The British Isles, Hydrographical Office, Admiralty, London 1832.
[52] The Fresnel Lens Makers – Thomas Tag, United States Lighthouse Society.
[53] The Fresnel Lens Makers – Thomas Tag, United States Lighthouse Society.
[54] The Lighthouse Work of Sir James Chance, Baronet – James Frederick Chance; Smith, Elder & Co., London 1902.
[55] The Lighthouse Work of Sir James Chance, Baronet – James Frederick Chance; Smith, Elder & Co., London 1902.
[56] Optical Apparatus as used in Lighthouses – James T Chance; William Clowes and Sons, London 1867
[57] Report from the Select Committee on Lighthouses- London 1845.
[58] Report from the Select Committee on Lighthouses- London 1845.
[59] A Rudimentary Treatise on the History, Construction and Illumination of Lighthouses – Alan Stevenson; John Weale, London, 1850.
[60] A Short Bright Flash – Theresa Levitt, W. W. Norton, New York, 2013.
[61] A Few Notes on Modern Lighthouse Practice – Chance Brother & Co. Ltd., Birmingham, 1910.
[62] A Description and List of the Lighthouses of the World – Alexander George Findlay; Richard Holmes Laurie, London, 1865.
[63] A Description and List of the Lighthouses of the World – Alexander George Findlay; Richard Holmes Laurie, London, 1865.
[64] https://trinityhousehistory.wordpress.com/tag/south-goodwin-lightvessel/

Chapter Nine
Warning by Sound

Prior to the final quarter of the nineteenth century, most cargo was still carried in sailing vessels, although steamships had taken most of the passenger trade. As we saw earlier, the problem was that the early steam engines were not very efficient and hence the ship had to carry so much fuel there was insufficient space left for enough cargo for the shipowner to carry it and make a profit. All that changed when engines that worked at a higher steam pressure and made better use of the steam became available. The energy from the more highly pressurized steam was used in three stages using high, intermediate and low-pressure cylinders. This type of engine was called the triple expansion engine and it was far more efficient. Steamships now needed far less fuel and could compete with sailing ships for the carriage of general cargo. Nevertheless, they still faced exactly the same navigational issues as did the sailing ships. Not only did this mean that Findlay's work retained its importance, it actually enhanced it. The steamships tended to be bigger, carry more cargo and hence their loss would be far more costly. Their engines might help them evade being driven ashore in a storm, but increased their exposure to another hazard.

A steamship could maintain speed without regard to the wind, but as they approached the land they were more vulnerable to another hazard of the climate – fog. While the captain of a sailing ship, just like his colleague in a steamship, was faced with all the problems inherent in reduced visibility, because fog is generally associated with calm weather and little if no wind, then his ship would only be moving very slowly, whereas the steamship could continue at full speed. The waters around the British Isles are susceptible to fog as are parts of the eastern seaboard of North America and anything that could be done to alleviate the dangers that fog posed would clearly be welcomed. In the nineteenth century the only alternative to marking a danger with light was to mark it with sound.

Robert Southey's famous poem, The Inchcape Rock, has it that one of the first audible warning devices was the bell supposedly erected on the Inchcape Rock by the Abbot of Aberbrothok and removed by Ralph the Rover. By the turn of the eighteenth century, the Inchcape or Bell Rock, was marked by Stevenson's famous wave-swept lighthouse, but even that could provide little protection in fog. However, by 1812 a bell had been attached to the balcony of the Bell Rock lighthouse and was sounded at regular intervals when visibility was low. Up to the year 1860 bells, weighing between 3 and 45 cwt.[65] were operated at various lighthouses around the coast. The bell itself did not move, but was struck by a clapper which was driven by a mechanism to ensure that it was sounded at regular intervals.[66] While these bells may have offered some warning, the distance at which they could be heard varied very considerably, largely depending upon the direction of the wind.

Sound waves travel by vibrating air particles so how far they travel will inevitably be greatly influenced by both any movement of the air and any change in the density of the air.

The consequence was that the distance at which the bell could be heard constantly varied. What did seem to make a difference was the size of the bell, bigger bells being audible at a greater distance. However, the size of the bell that you could attach to the top of a wave-swept lighthouse was obviously limited. An alternative was to use explosives, rather like big fireworks, the explosion producing a loud bang. While this might be heard at a greater distance, there was obviously a limit to the number of such devices that could be held at any lighthouse and so the intervals between the explosive signals had to be much longer.

A more effective and reliable instrument for the production of an intense sound, known as the siren is described in Findlay's 1899 List of Lights and Fog Signals. A siren, built by Professor Henry of the United States Lighthouse Board, was shipped to Britain in 1873 and tested by Dr. Tyndall at the South Foreland lighthouse.

> The principle of the siren is easily understood. A musical sound is produced when the tympanic membrane is struck periodically with sufficient rapidity. The production of the tympanic shocks by puffs of air was first realised by Dr Robison, and his device was the first and simplest form of the Siren. A stop-cock was so constructed that it opened and shut the passage of a pipe 720 times in a second. Air being allowed to pass intermittently along the pipe by the rotation of the cock, a musical sound was most smoothly uttered.[67]

As might be expected, further improvements soon followed, one of the most important being the addition of a very large cast iron trumpet, some 20 feet in length, which acted as a kind of loudspeaker. Just as the light emanating from a lighthouse had to be recognizable by a distinct characteristic, so too did a fog signal and hence there was every much as great a need for a List of Fog-Signals as there was for lights. The 1890 edition of Findlay's Lists acknowledges this.

> With the present EDITION of 'Lighthouses of the World' is incorporated a List and Description of Fog Signals, a comparatively recent aid to navigation, which has become a rapidly developing necessity by reason of the great increase of steam shipping in all parts of the world.

The information in the list of Fog Signals is presented in five columns as shown in Table 13.

Table 13		Fog Signals		
Name of Place	Lat. N Long. E	Signal	Period of Signal, and other Particulars	Year established
As is the case with lights, some of the Fog Signals are classified as major signals.	-	Bell Gong Siren Reed Horn Gun Rocket Trumpet Whistle	In the case of bells, the weight is often given together with how the bell was struck, e.g., by machinery or by hand.	-

Perhaps because the inclusion of fog signals was a recent change in the format of Findlay's Lists, a map the displaying the location of major fog signals in 1888 was incorporated in the 1890 Edition. On the map the location of fog signals was indicated by one of a number of circular symbols. Unfortunately, no key was provided and it is not possible in all cases to identify a particular symbol. Even where the symbols can be identified, when cross-referenced with the information on the lists there is a lack of consistency. However, as the map is dated 1888 and the list 1890, this may account for some of the discrepancies. A representation of the map is shown in Fig. 44.

Fig. 44   Location of Fog Signals by Type around the British Isles in 1888.

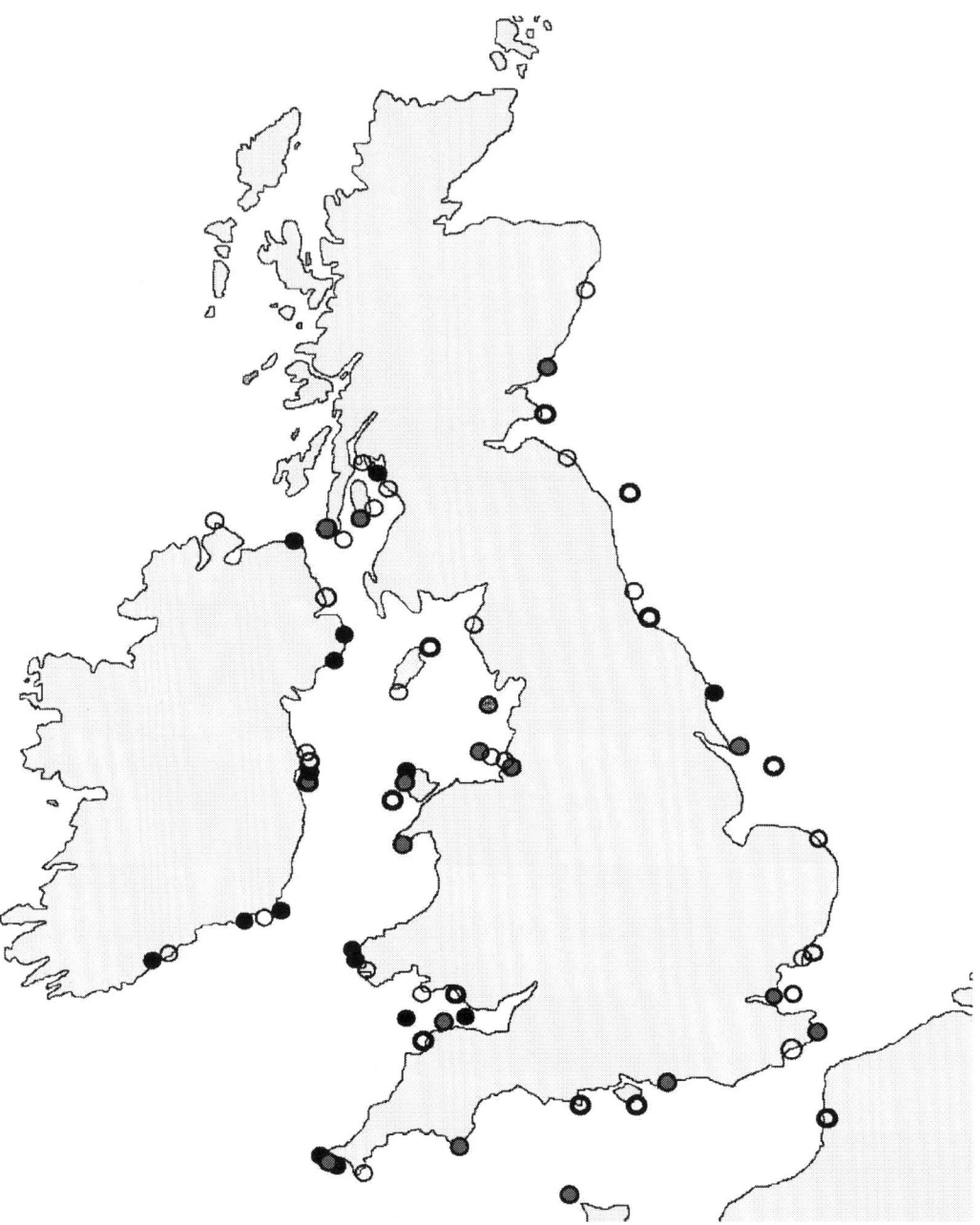

The map makes it evident that fog signals were concentrated in those places where vessels would approach close to land in order to reach a port, where they might first sight land or where there was some hazard relating to shallow water.

An analysis of the data contained in Findlay's Lists for the British Isles is shown in Table 14.

Table 14		Fog Signals by Type for the British Isles		
Year	Signal	Major Lights		Minor Lights	
		Land Based*	Light Vessels	Land Based*	Light Vessels
1890	Bell	-	-	53	4
	Gong	-	1	-	38
	Gun	-	4	12	1
	Reed Horn	4	7	-	-
	Rocket	1	-	1	-
	Siren	25	19	-	-
	Trumpet	2	-	1	-
	Whistle	4	-	-	-
1899	Bell	-	-	59	4
	Gong	-	-	7	7
	Gun	-	-	20	4
	Reed Horn	5	9	2	27
	Rocket	-	-	1	-
	Siren	38	23	-	-
	Trumpet	2	-	-	-
	Whistle	3	-	-	1
Note	*Includes all rock based and pile lighthouses.				

The data in Table 14 shows very clearly that the most favoured form of fog signal for major fog signal stations, whether land based or afloat, was the siren. Between 1890 and 1899 it is evident that there had been some standardization, with the use of gongs, guns and rockets being discontinued. In terms of minor land-based fog signal stations, bells continued to be the most favoured form of signal, but for the very considerable number of minor light vessels there had been a significant change from gongs to reed horns.

It should be noted that in 1880, 46% of the major fog signals were operated from light vessels, while in 1899 it was slightly lower at 40%. It might be asked why it was considered necessary to have so many fog signals out to sea. The answer is a twofold one. Firstly, without the opportunity to observe the shore and thereby establish an accurate position, shallow water represented a major hazard to shipping in fog. Secondly, the light vessel itself needed to ensure that ships knew where it was, for being of necessity anchored close to a shipping lane, it represented an obstacle that passing ships needed to avoid. The importance of the latter issue may be judged by the fact that between 1877 and 1900 there were fourteen collisions involving the Goodwin Light Vessel, fortunately without any loss of life.[68] It was even more starkly demonstrated in 1934 when the Nantucket Lightvessel off the eastern coast of the United States was rammed and sunk in thick fog by the liner RMS Olympic and regrettably on this occasion lives were lost.

In only two other areas of the world did Findlay's List show a significant number of fog-signals, Europe and Eastern North America. The data for these areas in shown in Table 15.

Table 15		Fog Signals by Type for Europe and Eastern North America.							
		Europe				Eastern North America			
Year	Signal	Major Land Based*	Floating	Minor Land Based*	Floating	Major Land Based*	Floating	Minor Land Based*	Floating
1890	Bell	-	-	78	32	-	-	148	19
	Gong	-	-	4	5	-	-	-	-
	Gun	-	-	13	2	2	-	8	1
	Reed Horn	8	5	5	2	20	-	1	1
	Rocket	1	-	-	-	-	-	-	-
	Siren	23	21	2	1	9	2	-	-
	Trumpet	9	-	1	1	17	-	-	-
	Whistle	-	4	-	-	32	7	-	-
1899	Bell	-	-	61	4	-	-	195	17
	Gong	-	-	7	7	-	-	-	-
	Gun	-	-	20	4	-	-	13	-
	Reed Horn	10	5	6	3	16	-	31	-
	Rocket	1	-	-	-	-	-	-	-
	Siren	45	39	3	-	16	-	1	-
	Trumpet	8	2	1	-	18	-	-	-
	Whistle	-	3	-	-	31	21	3	-
Note		*Includes all rock based and pile lighthouses.							

It will be noted that there is a significant difference between the two areas in the type of apparatus favoured for major fog signals. Around the British Isles and the coasts of Europe, the favoured instrument was the siren, whereas on the eastern seaboard of North America, it was the whistle. There does not seem to be any scientific evidence that would objectively establish which type of signal was superior and it may well have been no more than a matter of preference. There is a similar, although less marked, difference in the case of minor fog signals with Europe making much greater use of gongs and guns. Only one station, Heligoland, appears to have made use of rockets. However, while the use of bells as a fog warning signal was very widely employed in North America, the use of gongs was completely eschewed.

The 1899 Edition

In 1867 Findlay's slim volume had cost 3/6 (equivalent to £20); by 1890 the edition with the list of Fog Signals was considerably thicker and had more than doubled in price, selling for seven shillings and six pence, 7/6, the equivalent of £45. In 1899 the price had risen again to eight shillings, 8/-, the equivalent of £52, but now included an additional section, although it was far from being a cheap product. The additional section was a set of Tide Tables which was organized on an oceanic basis, preceded by a short discourse on why tides occur. The format of these tables does not provide any specific times for high or low water, but rather gives a single time by which the actual time of the tide can be calculated if the time of transit of the moon is known. Why the information is presented in this way is unclear, but as a conventional tide table would have entries for every day of the year, it may simply have been a way to economize on space.

The preface to the 1899 List provides a glimpse, not only of the changing world of navigation, but also of the impact of this and other factors on the world of publishing.

> The hitherto unexampled activity displayed by the home and foreign Lighthouse Authorities, during the last few years, has led to so great an increase in the number of coast and harbour Lights that it has again become necessary to considerably extend the List given in this Work. In the THIRTY-SEVENTH EDITION (1897), therefore, the Lists were again carefully revised, and the *List of Lights* extended from 224 to 305 pages, or almost *double* the number in the 1890 Edition, and a new Index was compiled.
>
> In this THIRTY-NINTH EDITION (1899), the *List of Fog Signals* has again had to be extended several pages, and a new Index compiled.
>
> The user of the List is assured that the information contained therein is as accurate as possible and that the official information is supplemented by practical experience.
>
> The Lists are corrected annually by the descriptions and notices issued by the home and foreign authorities; by personal observation; and by information kindly forwarded to us by commanders of vessels and others.

There is also a somewhat veiled indication that the publisher had begun to find it harder to make this particular publication profitable.

> Notwithstanding the great increase in the size of the Book, the price has been but slightly advanced, and the publisher, therefore, looks to those who have found it a reliable guide to extend its circulation, as some recompense

for the care and labour expended over it, and for the preparation and gratuitous supply of the Annual Supplements.

The last few pages are devoted to an abridged catalogue of the publications offered by Richard Holmes Laurie. They give an interesting insight into the cost of charts and other nautical publications at this time.

12   THE INDIAN AND PACIFIC OCEANS
From the Cape of Good Hope to Cape Horn; with plans of fifty-two of the principal Harbours. Arranged on twelve sheets of double elephant paper. By A. G. Findlay. F.R.G.S. Price for the whole … … £2 8 0

The quote above was for just one of the 200 charts advertised. In addition, there were Nautical Directories covering the Atlantic, Indian and Pacific Oceans by Alexander George Findlay, F.R.G.S.

3   THE INDIAN OCEAN
With descriptions of its Coasts, Islands, &c., from the Cape of Good Hope to the Strait of Sunda and Western Australia; including also, the Red Sea and Persian Gulf; the Winds, Monsoon and Currents, the Passages from Europe to its various Ports, and the Port Regulations and Charges; Indian Money, Weights and Measures, &c. With numerous illustrations. In one volume, royal octavo, 1,342 pages. *Fourth Edition 1882. Addenda 1889* … … … … … … £1 8 0

Alexander George Findlay died in 1875 and the 1899 edition of Findlay's Lists had been edited by William R. Kettle F.R.G.S.

[65] 'cwt.' is the recognized abbreviation for one Imperial hundredweight – one twentieth of an Imperial ton.
[66] A Description and List of the Lighthouses of the World – Alexander George Findlay; Richard Holmes Laurie, London, 1880.
[67] Sirens – E. Price Edwards, from A Description and List of the Lighthouses of the World – Alexander George Findlay; Richard Holmes Laurie, London, 1880.
[68] Lighthouses-The Race to Illuminate the World – Toby Chance and Peter Williams, New Holland, London, 2008.

Chapter Ten
The Last Voyage of the Margaret Smith

Greenock, situated on the south bank of the River Clyde close to where it meets the Firth of Clyde, was an important port in the nineteenth century. Today the riverside docks at Greenock are virtually devoid of commercial activity, they offer shelter to a few tugs and are partially used as a marina. As you walk around them they seem far too small to have ever offered berths to ocean going vessels, but that is because in the twenty-first century our experience tells us that ships are very much larger and even a cross-channel ferry would dwarf a ship like the *Margaret Smith*. However, in 1879 it was a very different story for the docks would have been full of sailing ships loading and discharging cargo.

Fig. 45 The Old Docks at Greenock.

East Harbour Greenock

Victoria Harbour Greenock

In common with every other British ship sailing to a foreign port, the crew of the *Margaret Smith* were employed on the basis of a legal document called a 'Crew Agreement.' Most of the information which follows is drawn from this document, the original of which is held by the Maritime History Archive at the Memorial University, St Johns, Newfoundland, Canada. The agreement was a printed document that had to be filled in before every voyage and was carried on the ship as part of its official papers. It had to be returned to the Mercantile Marine Office in London at the end of the voyage.

Most of the information had to be filled in by hand and while some of this was probably done by a clerk who had very neat handwriting, their spelling and punctuation differs somewhat from modern English. However, much of the information, especially the names of the crew, was written in by the men themselves if they were literate. In some cases, their writing was very poor and hence is rather difficult to read, particularly as they did not always confine what they wrote to the allocated space. A further difficulty is that the document was inevitably handled by seamen whose hands were probably less than clean and so the document has a somewhat dirty and well-worn appearance. All of these factors make some of the entries difficult to interpret. Nevertheless, it has been possible to extract most of the information.

When the *Margaret Smith* left Greenock in July 1879, she carried a crew of 15 as listed in Table 16.

Table 16	The Crew of the *Margaret Smith* - July 1879		
Name	Position	Age	Birthplace
Alexander Taylor	Captain	49	Argyllshire, Scotland
Daniel McAllister	Mate	41	Greenock, Scotland
Edmund Darragh	Boatswain	23	Belfast
John McAuslan	Carpenter	*	Rothsay, Scotland
George Wylie	Cook	57	Greenock, Scotland
Francis Renand	Steward	21	Mauritius
Alexander Currie (X)	Able Seaman	49	Greenock, Scotland
Matthew Collope (X)	Able Seaman	29	Trieste
Antonio Johnson	Able Seaman	29	Trieste
James McGlinchy	Able Seaman	23	Greenock, Scotland
Alexander Bohlin	Able Seaman	29	Sweden
Albert Murray	Able Seaman	23	Mauritius
William Green	Able Seaman	23	Mauritius
Isaac Davies	Ordinary Seaman	16	Greenock, Scotland
Chas. Darragh	Apprentice	17	Greenock, Scotland
Note	(X) indicates that the man was illiterate and made his mark X, the name being written in by someone else. * Overwritten by the man's name and cannot be transcribed with confidence.		

The crew of any ship were required to 'sign-on'. This meant that they had to sign the 'Crew Agreement' which set out the terms and conditions under which they would be employed. Thus, it specified such things as where the crew might be obliged to take the ship, what food would be provided for them, how much they would be paid and any other special conditions. It also specified the maximum duration of the agreement, which in this case was three years.

While the first stage of the voyage was to be from Greenock to Mauritius, the agreement allowed the shipowner the opportunity to continue from Mauritius to more or less any other part of the world. Effectively, the voyage did not end until the vessel returned to the United Kingdom. In the following extract the italicized portions represent the hand-written parts.

> The several Persons whose names are hereto subscribed, and whose descriptions are contained on the other side or sides, and of whom *seven* are engaged as Sailors, hereby agree to serve on board the said Ship, in the several capacities expressed against their respective Names on a voyage from *Greenock to Mauritius and if required to any port or dock in the*

Indian or China Seas and North or South Pacific Oceans and Australian New Zealand or Cape Colonies and East or West Coasts of North or South America and North or South Atlantic Oceans and West Indies and United States of America and British North America and Mediterranean Sea and any part of Europe to and from as freight or employment may offer for the ship and until the return to a final port of discharge in the United Kingdom or Continent of Europe with liberty to call at a port or ports for orders.
Term of Service not to exceed three years.

There were a considerable number of other standard elements to the crew agreement whereby the crew agreed to obey the lawful commands of the ship's master and diligently carry out the duties assigned to them. In exchange, the shipowner agreed to provide the crew with provisions and pay them the wages on the basis clearly set out in the agreement. It was also stipulated that:

Any Embezzlement or wilful or negligent Destruction of any part of the Ship's Cargo or Stores shall be made good to the owner out of the Wages of the Person guilty of the same: And if any Person enters himself as qualified for a duty which he proves incompetent to perform, his Wages shall be reduced in proportion to his incompetency:

Apart from the usual legal definitions and explanations, there were some additional agreements specific to this particular voyage.

There shall be no right to any allowance of spirits no liberty to go on shore nor any money advanced in a foreign port.

In effect, the crew could be required to remain on board the ship for up to three years without receiving any of the pay due to them, conditions which would seem to be fairly onerous. The agreement set out the pay each man would receive. This is shown in Table 17.

Table 17	Pay for the Crew of the *Margaret Smith*			
Name	Position	Pay per Month	Pay Advanced	Monthly Allotments
Alexander Taylor	Captain	-	-	-
Daniel McAllister	Mate	£6 - 0 - 0	£6 - 0 - 0	£3 - 10 - 0
Edmund Darragh	Boatswain	£3 - 10 - 0	£3 - 10 - 0	£1 - 10 - 0
John McAuslan	Carpenter	£5 - 0 - 0	£5 - 0 - 0	£1 - 10 - 0
George Wylie	Cook	£2 - 10 - 0	£2 - 10 - 0	£1 - 10 - 0
Francis Renand	Steward	£2 - 15 - 0	£2 - 15 - 0	£1 - 10 - 0
Alexander Currie (X)	Able Seaman	£2 - 0 - 0	£2 - 0 - 0	£1 - 0 - 0
Matthew Collope (X)		£2 - 0 - 0	£2 - 10 - 0	-
Antonio Johnson		£2 - 0 - 0	£2 - 10 - 0	-
James McGlinchy		£2 - 0 - 0	£2 - 0 - 0	-
Alexander Bohlin		£2 - 0 - 0	£2 - 0 - 0	-
Albert Murray		£2 - 0 - 0	£2 - 0 - 0	-
William Green		£2 - 0 - 0	£2 - 0 - 0	-
Isaac Davies	Ordinary Seaman	£0 - 15 - 0	£0 - 15 - 0	-
Chas. Darragh	Apprentice	-	-	-
Note	(X) The man was illiterate and signed with a mark, the name being written in by someone else.			

It is interesting to note from Table 17 that the ship's carpenter was paid almost as much as the Mate, a clear indication of the importance attached to his role. Although there is no specific explanation, it might reasonably be assumed that the monthly allotments noted in Table 17 were sums that supported the families of the men while they were at sea. It is also noteworthy that the pay offered in 1879 was in many cases lower than that in 1869. The Mate received an additional ten shillings (£0 – 10 – 0) per month, as did the carpenter and the boatswain's pay was unaltered, but the pay of everyone else was between five shillings and fifteen shillings lower, with the able seamen seeing the biggest reduction, with their pay reduced by around a quarter.

It is evident from Table 16 that the crew came from a number of different backgrounds. Seamen frequently signed on for a new voyage shortly after returning from a previous one. Thus, it was common that a ship would have a largely different crew on every voyage. The previous employment of the men who formed the crew of the *Margaret Smith* is shown in Table 18.

Table 18	Previous Employment of the Crew of the *Margaret Smith*		
Name	Position	Previous Ship	Date and Place of joining this Ship
Alexander Taylor	Captain	Same Ship	continued
Daniel McAllister	Mate	Oriana, London	24/7/79 Greenock
Edmund Darragh	Boatswain	Margaret Smith, Greenock	
John McAuslan	Carpenter	Culzean, Greenock	
George Wylie	Cook	Greenock, Greenock	
Francis Renand	Steward	*	
Alexander Currie (X)	Able Seaman	Greenock, Greenock	
Matthew Collope (X)		Madras, Port Glasgow	
Antonio Johnson		Demerara, Greenock	
James McGlinchy		Cambria, Newquay	
Alexander Bohlin		Souvenir, Yarmouth, Nova Scotia	
Albert Murray		Morayshire, Glasgow	
William Green		Abbey, Greenock	
Isaac Davies	Ordinary Seaman	Challenge, Greenock	
Chas. Darragh	Apprentice	-	
Note	* Neither the name of the ship or the port can be transcribed with confidence.		

Tuesday the 29th July 1879 would have been a hectic day for the crew of the *Margaret Smith* as they made their final preparations for the long voyage to Mauritius, the 48th voyage that the ship had undertaken. Theirs was not a new ship, she had been launched from the Albert Quay shipyard of Messrs James McMillan and Son of Greenock on the 4th September 1857, the launching ceremony being performed by Miss McMillan. Named after the daughter of her owner, John Kerr of Greenock, she was 158 feet long, 29½ feet in breadth and 19 feet in depth with a registered tonnage of 650 tons. She was not the first of John Kerr's ships to bear his daughter's name. Her predecessor, a much smaller barque of 258 tons, built in Greenock in 1851[69] had been wrecked, fortunately without loss of life, on Inagua Island in the Bahamas in January that year.[70] As this new *Margaret Smith* was launched, she collided with the *Alberto*, a timber laden vessel being towed upriver. However, while the *Alberto* lost her main yard and the *Margaret Smith* damaged her rudder, no serious damage was done.[71]

In 1879 John Kerr operated a fleet of eighteen sailing ships, most being what was called 'ship rigged' and the rest barques. They were largely engaged in the sugar trade from the West Indies and modern-day Guyana. The Kerr fleet in 1879 is shown in Table 19.

Table 19				The Kerr Fleet in 1879[72]
Name of Vessel	Type	Tonnage	Captain	Trade in which Engaged
Barbadian	Ship	700	McMillan	Foreign
Bargany		1,232	Taylor	
Berbice		717	Quail	
Cochin		1,200	Oudney	
Culzean		1,372	Pirnie	
Demerera	Barque	486	McNeill	
Greenock		840	Manson	
Killochan	Ship	1,233	McFadzean	Coastal
Kilberan		1,199	Lowe	Foreign
Kurrachee		695	Masson	
Lapenstrath	Barque	623	Logan	
Madras		668	Murdoch	
Margaret Smith		631	Taylor	
Mauritius	Ship	682	Ferguson	
Orissa		1,199	Maxwell	
Trinidad		772	Manson	
Trochbrague	Barque	677	Thomson	

The *Margaret Smith* had, as we have already seen, endured her fair share of storms, suffering considerable damage, but this had always been properly repaired and the vessel returned to service. Twenty years old in 1879, she had completed 47 voyages and sailed well over 430,000 nautical miles.[73] She had, however, suffered a more serious mishap in March 1877 when she ran aground on Diamond Rock in the Grand Bocas, shortly after leaving Trinidad with a cargo of sugar for Greenock.[74] The crew were able to get her off the rock after three hours, but she had to return to Trinidad where her cargo was unloaded and transferred to another of John Kerr's ships. A surveyor, Mr. Wimshurst, examined the damage on behalf of the underwriters and recommended that the *Margaret Smith* be towed to Martinique[75] where she arrived on the 26th May 1877.[76] The repairs cost 30,000 francs and took until July when she sailed from Trinidad to Cardiff under the command of Captain Murdoch.[77]

In July 1879, the Margaret Smith's master, Captain Alexander Taylor and First Mate Daniel McAllister should have ensured that the ship was in all respects ready for sea, manned by a competent crew and fully stored for the voyage. The Captain should also have ensured that he had available all the requisite instruments and information to navigate the ship on her 8,426 mile voyage.[78] The cargo of coal would have been loaded, the hatches secured and the necessary stores taken aboard, although almost certainly perishable items such as food would have been left to last possible moment. The ship carried a crew of fifteen, a captain, mate, boatswain, carpenter, cook, steward, seven able seamen, an ordinary seaman, and an apprentice. Each man would have 'signed on' for the voyage. As noted previously, the 'Crew Agreement' document was very specific about the food to be supplied to the men. Table 20 shows what they could expect each day.[79]

Table 20							Scale of Provisions	
Day	Bread lb.*	Beef lb.	Pork lb.	Flour lb.	Peas pints	Tea oz.**	Coffee oz.	Water qts.***
Sunday	1	1½	-	½	-	¼	½	3
Monday	1	-	1½	-	1/3	¼	½	3
Tuesday	1	1½	-	½	-	¼	½	3
Wednesday	1	-	1½	-	1/3	¼	½	3
Thursday	1	1½	-	½	-	¼	½	3
Friday	1	-	1½	-	1/3	¼	½	3
Saturday	1	1½	-	-	-	¼	½	3
Note	*abbreviation for one pound (0.45 kg). **abbreviation for an ounce. There were 16 ounces to the pound. ***abbreviation for quarts. A quart was two pints and 8 pints made an Imperial Gallon (4.55 litres).							

It is worth noting that for this voyage the size of the crew had been substantially reduced from the 28 carried on a voyage in 1869, with the number of seamen being halved. The scale of provision was also rather less generous than that provided ten years earlier in that the weekly issue of 2 lbs. of rice is no longer listed and the quantity of tea is halved.

The voyage to Mauritius would take over three months, so if all the provisions had to be taken on board in Greenock the quantities would have been very large. Unless the ship was able to re-provision during the voyage then the minimum quantities would be as shown in Table 21.

Table 21	Minimum Quantity Provisions for the *Margaret Smith*	
Item	Quantity in lbs	Equivalent in kgs
Bread, presumably in the form of flour	1,575	709
Beef	1,350	608
Pork	1,013	456
Flour	338	152
Peas	151	68
Total	**4,427**	**1,993**

However, if food was a major item, then water was an even bigger issue. An allowance of 3 quarts (3.4 litres) per man per day meant that with a crew of 15 men, they would require 315 quarts per week or 79 Imperial gallons (359 litres). Over the voyage this would mount up to around 1,185 Imperial gallons (5,387 litres), roughly 11,850 lbs. (5,375 kg) or over 5 tons and it is unlikely that the water would have stayed palatable for such a long period of time. Therefore, calling at a port for fresh water would be more or less essential. This would mean that the ship would have to approach the coast and locate a suitable harbour, something for which Findlay's List of Lights would obviously be very useful.

It may well be that the crew of the *Margaret Smith* made the most of their last opportunity to say good bye to friends and family, or just enjoy a night out, but they must have risen very early on the Wednesday morning as by 8am the ship had been towed to a point just south of Ailsa Craig by the Clyde Shipping Company's *Flying Dutchman*.[80] There is no readily available information as to what ports the ship called, but we do know that she was sighted on the 17th August 1879 at a position 39N 17W, that is about 100 miles west of the coast of Portugal and that all was well.[81] The sighting of the *Margaret Smith* in this position would indicate that Captain Taylor was taking advantage of the Canaries current as shown in Fig. 11.

Three months later, by the 19th November, the island of Mauritius was in sight. It is not hard to imagine the relief of the crew as they realized that the long voyage was almost over. Captain Taylor had never sailed to Mauritius before, but First Mate McAllister had made two previous visits. Their final destination was Port Louis on the west side of the island, but when they made their landfall, it was on the east side, so they would have to sail either round the south or north to reach their destination. Captain Taylor decided that the northern route was the more appropriate and set a course accordingly. There are two islands to the north of Mauritius, Flat Island and Gunner's Quoin, the former being marked by a lighthouse, a 53 foot high tower, which showed a bright, revolving light with a period of one minute and which was visible at up to 25 miles.[82] Findlay's Directory for the Indian Ocean contained very specific instructions as to how a safe approach to Port Louis should be achieved.

Directions. – The following are the official instructions issued by the Mauritius authorities. They are based upon those given by Mr. Kelly, harbour master, in 1855, and by Mr. Wales, in 1858.

Vessels arriving from the eastward or south-eastward should be careful not to bring up the Light on Flat Island to the northward of N.N.W.½W. until Gunner's Quoin bears West, when they may pass midway between it and Flat Island; this course will lead about 2¼ miles clear of the reefs that extend from the N.E. end of Mauritius.[83]

Fig. 46   The Approach to Port Louis round the North of the Island of Mauritius.

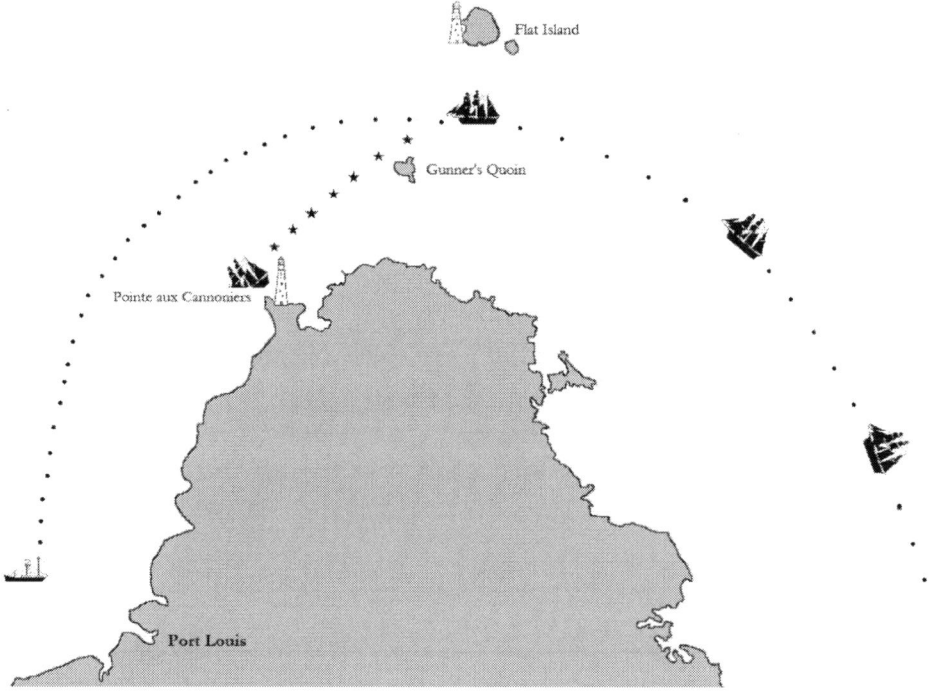

When to the westward of the Quoin, Cannonier Point light will be seen. Steer with the Flat Island light astern, bearing N.E. by E.½E., until the Cannonier Point light bears S.E. by S. (which will carry you clear of the dangerous reef that extends from the point): you may then haul up S.W. by S. till the red light at Grand River is seen.

There are then a whole series of instructions for smaller vessels that sail closer to the coast, but the directions then include the following information for larger vessels. It is noted that the lighthouse on Cannonier Point, which normally shows a fixed, bright white light, changes to a red light inshore of S.W. by W.

It is not well, however, for a large vessel ever to approach so near to the reef as to change the Cannonier Point light from white to red. The great object in view in thus arranging the light was the convenience of the numerous coasting vessels belonging to the colony, to whom (knowing the ground as they do) it is most useful. The best mark for keeping clear of the reefs between Cannonier Point and Point Piment by night is to keep the Flat Island light open to the westward of the Cannonier Point light until the red light at Grand River is seen.

The correct course is illustrated by the dotted line in Fig. 46. Since the 1866 directions were published, a light vessel had been stationed just to the north of the entrance to Port Louis, marking the safe anchorage. This light vessel exhibited a bright, revolving light with a period of half a minute and was visible at a distance of nine miles.

The weather was clear and the sea calm that night, visibility was good, so there should have been no difficulty in bringing the *Margaret Smith* to the safe anchorage off Port Louis. However, as Fig. 46 shows, Captain Taylor put his ship on a course that led directly to the reef at Cannonier Point where she struck. It was quickly apparent that there was no hope of getting her off the reef and that she was a wreck. The only good thing being that no lives were lost. The Greenock Advertiser of the 20th December 1879 carried the following brief announcement.

> A Lloyd's telegram, dated Aden, 19th Dec., says the *Margaret Smith* (British barque), of Greenock, got ashore at Cannoniers Point on 20th November, and is a total loss. Nothing saved. Vessel and cargo sold, realising £648; materials £33. The master's certificate has been suspended for twelve months.

What had happened to allow this disaster to happen? This was the question that the Mauritius Marine Board set out to answer. The following extract from the Minutes of the Board leaves no doubt as to what actually occurred.[84] The Board had already ascertained that the *Margaret Smith* was tight, staunch, seaworthy and well equipped, so the most likely cause would be human error.

> The Master then stood up along the land, and passed between Flat island and Gunner's Quoin at about 8pm. Up to this time, the master who had never previously been to Mauritius, appears to have navigated the vessel according to his own ideas, but after passing between the islands he appears to have been guided in great measure by the opinion of his Chief Officer, who had already made two previous voyages to Mauritius in that capacity. It

appears that on rounding the Quoin a light was seen on the port bow, and the vessel was steered towards it, the Chief Officer stating to the master that it was the lightvessel, and not discovering his mistake until a few seconds before the vessel grounded on the reefs to the north of Cannonier's Point, and became a total wreck. A cast of the lead was taken by the Master a few minutes before she struck, and the vessel was found to be in 4½ fathoms, from which time the Master commenced to shorten sail in order to come to an anchor, being evidently up to the last moment under the impression that he was standing for the lightvessel, instead of the lighthouse on Cannonier's Point. As soon as the vessel struck the master ordered the small boat to be got out and proceeded with two men in search of assistance, leaving orders to the Chief Officer to do his best to get the vessel off. No serious efforts, however, seem to have been made beyond backing the sails, and it appears from the evidence that it would have been useless, as the vessel could not have floated; the rudder was unshipped the second time she bumped, and the first quarter of an hour the water gained 9 inches in the well, gaining rapidly on the pumps during the night, and at 7am on the 21st there were 9 feet 6 inches in the hold.

The foregoing quote from the Board's minutes establishes in considerable detail what actually happened and it is clear that there was no hope for the *Margaret Smith* once she had struck the reef. The Board's minutes go on to provide further evidence about how the ship came to be in such a position.

There was slight commanding breeze, a weather shore, two first-class lights, one very close, a tolerably clear night and a moon; these ought to have been conditions of almost absolute safety, but, as before stated, the Master appears to have trusted almost entirely to his Chief Officer, not even to have taken the trouble to consult the book with which he was provided, entitled "Lights of the World" which gives a minute description of the Lights in the island of Mauritius. It seems hardly credible that the Master of a vessel coming to Mauritius for the first time should have neglected to take every opportunity of gaining information respecting the lights of the coast, anchorage, &c., especially in the absence of a chart of the island on a large scale; such, however, seems to have been the case. The ignorance and indifference of the Chief Officer are even more deplorable, for, having already made two previous voyages to Mauritius, and by his own evidence having laid on both occasions close to the lightvessel, he did not even remember whether the light shown by the lightvessel was fixed or revolving, and had taken no pains to refresh his memory.

There is absolutely no ambiguity in the Board's conclusions, the loss of the *Margaret Smith* was solely due to the negligence of her Master and Chief Officer. Reading the foregoing must have been a very sobering experience for Captain Taylor and Chief Officer McAllister, but much worse was to come.

> The Board, having carefully considered the case, are of opinion that the loss of the *Margaret Smith* was due to the careless and ignorant manner in which the vessel was navigated by the Master after passing Gunner's Quoin, in trusting entirely to the judgement of his Chief Officer on a coast of which he himself was utterly ignorant, and also in neglecting to use the lead carefully, which of itself would have given him an idea of his proximity to the land. The Master appears to have been wanting in a proper appreciation of his duties and responsibility as a Shipmaster from the moment he undertook the navigation of the *Margaret Smith* from Greenock until she was wrecked on the reefs. 1. He embarked without a large scale chart of Mauritius or books of instruction regarding the Indian Ocean, which he could have procured at nominal cost. 2. Admitting that they were forgotten, he had 90 days to read up and avail himself of sufficient information from the charts and books ("A Description and List of the Lighthouses of the World" 1878, by Alexander George Findlay, F.R.G.S.)[85] which he had with him, whereas he admits that he simply glanced at the page containing the necessary information. 3. In absolute ignorance of the coast he was approaching, he presumed recklessly to sail along it, and seek the anchorage during the night, instead of laying off until daylight. The conduct of the Chief Officer, he having made two previous voyages to this port, and being so unobservant as not to remember whether the lightvessel showed a fixed or revolving light, denotes a hopeless lack of intelligence in a man who aspires to the responsibility of navigating a ship with lives and property over the ocean.

The verdict of the Board could not have been clearer, the loss of the *Margaret Smith* was entirely attributable to the carelessness and recklessness of Captain Taylor and Chief Officer McAllister. They must, therefore, have anticipated that the Board's recommendations would be very uncomfortable, an anticipation that was quickly confirmed.

> We, therefore, recommend that the certificate of the Master, Mr Alexander Taylor, be suspended for a term of 12 months from the date of this meeting, and that of the First Mate, Mr Daniel McAllister, for 6 months from date, subject always to the approval of his Excellency the Governor of this colony

and the ratification of the Board of Trade. This decision was pronounced in open Court in accordance to the third clause of the 254 Section of the Merchant Shipping Amendment Act of 1862.

The punishment might seem fairly mild, but it would probably have signalled the end of two careers. The loss of the *Margaret Smith* and the subsequent inquiry, although reported in the local newspapers, were accorded only a scant space where the bare facts were given without comment. Nevertheless, it would undoubtedly have been a topic of much discussion in Greenock.

However, the Marine Board's verdict carries with it a crucial message for all who navigated ships across the oceans. Findlay's work is an authoritative source of information, absolutely vital for the safe passage of any vessel. Navigators are expected not only to have a copy of Findlay's work readily available, but to have studied it carefully. Failure to do so was clearly considered as serious negligence.

If the Master and Mate found themselves in some difficulties after the loss of the ship, the crew faced problems of their own. When the *Margaret Smith* struck the rocks off Cannoniers Point on the 19th December 1879, then the voyage was deemed to have ended and with it the employment of her crew, hence their pay was stopped from that moment onwards. They were paid the wages owed to them up to the time when the ship was wrecked, in Port Louis, Mauritius, under the supervision of the Deputy Marine Superintendent and apparently in Rupees as shown in Table 22.

Table 22 Final Payment to the Crew of the *Margaret Smith*	
Name	**Wages in Rupees**
Daniel McAllister	R 24.60
Edmund Darragh	R 30.27
John McAuslan	R 79.69
George Wylie	No wages due
Francis Renand	R 20.29
Alexander Currie (X)	R 14.34
Matthew Collope (X)	R 37.36
Antonio Johnson	R 49.96
James McGlinchy	R 40.73
Alexander Bohlin	R 57.34
Albert Murray	R 62.73
William Green	R 65.49
Isaac Davies	R 25.21

There are a number of factors in Table 22 which unfortunately cannot be explained. Daniel McAllister, the Mate and the highest paid member of the crew, received only R 24.60

which is considerably less than some of the seamen received. The cook, George Wylie, was, according to the information shown in Table 17 comparatively well paid, but for some unexplained reason is apparently due no wages at all. The lower payment to Francis Renand and Alexander Currie is probably due to the fact that there was a monthly allotment taken from their wages, but there is no explanation of why the wages of the other seamen are all different.

There is no readily available information about what happened to any of these men, other than that the Master Alexander Taylor and the Mate Daniel McAllister had their certificates suspended. Presumably, if they or any of the crew members wished to return to Greenock, then they would have had to sign on as part of the crew of a ship bound for a port in the United Kingdom.

[69] Lloyd's Register of British & Foreign Shipping – London 1856.
[70] Greenock Advertiser March 6th 1857.
[71] Greenock Telegraph 5th September 1857.
[72] Post Office Directory for Greenock 1879.
[73] Data drawn from Greenock Newspapers, Lloyd's List and www.https://sea-distances.org/ accessed 2019.
[74] Lloyd's List 13th April 1877.
[75] Lloyd's List 5th May 1877.
[76] Lloyd's List 29th May 1877.
[77] Lloyd's List 15th August 1877.
[78] www.https://sea-distances.org/ accessed April 2020.
[79] Crew Agreement for the *Margaret Smith* 1863.
[80] Greenock Advertiser July 30th 1879.
[81] Lloyd's List 26th August 1879.
[82] A Description and List of the Lighthouses of the World - – Alexander George Findlay; Richard Holmes Laurie, London, 1879.
[83] A Directory for the Navigation of the Indian Ocean with Descriptions of its Coasts Islands etc.- Alexander George Findlay; Richard Holmes Laurie, London, 1866.
[84] Lloyd's List 30th January 1880.
[85] Captain Taylor's version of Findlay's List was a year out of date but could have been updated by the supplements offered. However, as all the lights on Mauritius had been established some years earlier the information contained in the 1878 list would have been correct.

Chapter Eleven
Alexander George Findlay

Hitherto, we have looked at Findlay's work in relation to lighthouses in some detail and have briefly noted his work on Directories for mariners. However, to portray a fuller understanding of the man we need to look at his life and some of his other achievements. He was born in London in January 1812, the son of Alexander Findlay, also a geographer who was born in London in 1790. Alexander Findlay Senior was one of the original Fellows of the Royal Geographical Society when it was formed in 1830. The family was originally from Arbroath in Forfarshire, where Alexander George's grandfather had been a shipowner.[86]

The first area in which the young Alexander George Findlay made his name and built his reputation was in the compilation of Atlases of Ancient and Comparative Geography, but he soon began to move on to subjects more closely associated with matters marine. In 1851 he published the first of his famous 'Directories' which covered the coasts and islands of the Pacific Ocean. It consisted of 1,400 pages in two volumes and involved a prodigious amount of work. It has to be remembered that in those distant days everything had to be done on paper and the papers had to be kept in a filing system that was so organized that any part of it could be found with ease. The information came from a myriad of sources, often from mariners returning from voyages and so the information evolved continuously. Findlay must, of necessity, have been someone possessing of a very high level of organizational skills to have been able to produce reference works of such reliability that their users spoke highly of them. They were accepted as standard authorities in every quarter of the globe.[87]

An early interest in meteorology brought Findlay to the attention of Admiral Fitzroy who was sufficiently impressed as to invite him to join an official department then to be established. However, in a decision that was to shape the rest of his career, Findlay chose an independent path. He was already established as a respected authority on many matters relating to maritime affairs and in particular hydrography. In 1850 he married Sarah Rutley from Aylesford in Kent and according to the 1851 Census, he was living in Finsbury and working as a hydrographer. Already he had developed a deep interest on lighthouses, delivering papers on the subject to the Society of Arts in 1847 and 1858. These papers demonstrate not only a very thorough grasp of the basic technicalities of the subject, but also a keen appreciation of the importance of lighthouses in facilitating navigation and improving safety. Moreover, they were considered to be so important that the Society of Arts awarded him their medal.

The 1861 Census shows Alexander George and his wife Sarah living with his now elderly parents in Hayes, together with two children Helen aged 13 and William R aged 3 with the surname Kettle, the grandchildren of Alexander Senior, but clearly not the offspring of Alexander and Sarah. Indeed, there seems to be no record of their having had any children. Alexander George must have been very heavily involved with his work in the years leading up to 1861 for the first edition of his Lighthouse Lists was published that year. He had by this time added charts, directories, sailing directions for much of Europe, the Mediterranean, the Caribbean and eastern North America to his range of publications. It has been estimated that he had as many as 10,000 pages of highly detailed information to his name. Moreover, this information was constantly evolving and even though he must certainly have had some staff to assist him, the number of editorial decisions he would have had to make would have been very substantial. He was recognized as the foremost authority in this particular branch of nautical research and authorship.

Up to 1858 his work had been published by the firm of Richard Holmes Laurie, but Laurie died in 1858 and Findlay took over the business, something which must have added substantially to his workload. However, despite his heavy workload Findlay had always expressed a deep interest in matters outside the sphere of published materials. He worked on all the information concerning the possible routes of Sir John Franklin's ill-fated expedition and aided the Royal Geographical Society prepare the case for a further expedition in 1875. He was a friend of the African explorer Dr. Livingstone and devoted much time to investigating the questions surrounding the source of the Nile. He also found time to serve on various committees established by the British Association for the Advancement of Science, contributing especially to work on ocean currents. Alexander George Findlay was elected as a Fellow of the Royal Geographical Society in 1844, an honour of which it is evident he was very proud as the letters F.R.G.S. always appear after his name. His reputation spread abroad and in 1870 he was elected an honorary member of the Societa Geographica Italiana.

The 1871 Census shows Alexander George Findlay, hydrographer, living with his wife Sarah in Camberwell. He died in Dover, Kent on the 3rd May 1875, but his legacy of publications lived on. Others such as E. Price Edwards who wrote an introductory account for the 1880 edition of the List of Lighthouses, contributed to the contents and clearly the editorial work was undertaken by others whose names went largely unrecorded; it was the name of the originator, Alexander George Findlay F.R.G.S. that mattered. However, it may be recalled that the 1899 edition of Findlay's Lists had been edited by William R. Kettle F.R.G.S. and that in 1861 a three year old boy, William R Kettle, Alexander George's nephew had shared the same residence. It is difficult not to make the connection that the work was being kept in the family.

All the foregoing gives us a clear picture of Alexander George Findlay as a man who was not only professionally interested in his subject but extended that interest on a human scale. Today, much of his work has lapsed into obscurity, the lighthouses about which he was so passionate have largely been superseded by satellite navigation systems, radars and a host of other electronic aids. Information is available in digital format and updated at the touch of a button. Nevertheless, Findlay's work laid the foundation upon which others could build and for that we should be grateful.

During the research for this book the name of Alexander George Findlay appeared in a variety of places and contexts associated with charts, lighthouses and sailing instructions. It appeared in the Minutes of the Mauritius Marine Board and led to a much better appreciation of the contemporary importance attached to his work. However, it also came in a totally unexpected context.

The name of Alexander George Findlay was listed in a reference to a book called The Greatest Lie on Earth by Edward Hendrie published in 2016.[88] This volume is directly related to the work of Samuel Birley Rowbotham entitled Zetetic Astronomy, first published in 1865[88]. The sub-titles of these books indicate very clearly what they seek to demonstrate, for Hendrie's work, The Greatest Lie, is sub-titled Proof that Our World is Not a Moving Globe, while Rowbotham's Zetetic Astronomy' is sub-titled Earth Not a Globe. It is evident that while both these authors believe in the theory of a flat earth, Findlay was clearly convinced that the Earth was basically a sphere as he gave the positions of every light in terms of latitude and longitude, so it is difficult to see why they would quote his work. It appears that they were very ready to accept that Findlay was a recognized authority as the following quote demonstrates.

> He (Rowbotham) cites many instances in the authoritative text of its time on lighthouses, *Lighthouses of the World*, by Alexander G. Findlay (1861). Findlay was an internationally renowned geographer and hydrographer, who wrote many reference guides, which were invaluable to mariners throughout the world. Findlay had unquestioned competence in hydrography. Mariners relied upon the accuracy of his *Lighthouses of the World* reference guide. Indeed, it was imperative that the entries in that reference book be precise, as the very lives of the mariners rested on the accuracy of the information in that text.

It transpires that they were perfectly prepared to accept the validity of the data in Findlay's List as they considered it to provide incontrovertible proof of their assertion that the Earth was flat. In their view Findlay's data was beyond reproach, it was the interpretation that people put upon it that was at fault. It might be suspected that Findlay would not have agreed.

---

[86] Dictionary of National Biography.
[87] Royal Geographical Society Journal, Vol xlv 1875.
[88] The Greatest Lie on Earth – Edward Hendrie, Great Mountain Publishing, 2016.
[89] Zetetic Astronomy – Samuel Birley Rowbotham, Pantianos Classics.

EPILOGUE

As the twentieth century dawned, the need for lighthouses to aid navigation continued to grow. Ships were bigger and faster but still had to rely on the same navigational principles as they had in the nineteenth century. The development of the lighthouse remained a key priority. Lights needed to be as intense as possible to provide the maximum possible warning. The London International Lighthouse Conference of 1929 followed similar events in London 1923 and Cairo 1926. Key agenda items for the 1929 Conference were lighthouse illuminants and light intensities, the former paying particular attention to the increasing use of electric filament lamps.[90]

Fig. 47   Large Filament Lamp used in Lighthouses.

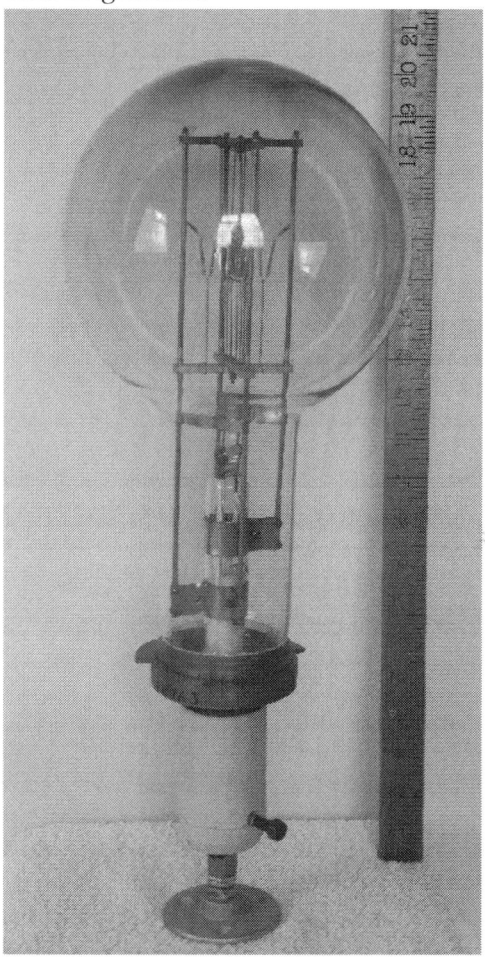

Giant Edison Screw Filament Lamp some 20 inches tall, from an English Lighthouse - Rated at 100 volts 3500 watts. (Photo Chris Hills)

Another conference was held in Paris in 1933 where it was agreed that the next venue for the conference in 1937, would be Berlin. Technical development required this conference to operate as three committees dealing with lighting, sound signals and wireless signals respectively.[91] The advent of wireless as an aid to navigation was clearly becoming an issue, but light and sound signals continued to be predominant. The next conference was scheduled for Holland in 1941, but was overtaken by the onset of war.

If wireless aids had been in their infancy in 1937, the needs of war dramatically accelerated their development and a critical area was finding a way to locate the submarines that were preying on the Atlantic convoys during the dark winter nights. Radar had been in existence since the late 1930's but by 1945, progress that might in peacetime had taken decades, had been such that ships could be fitted with equipment that used a cathode ray tube to present a picture of their surroundings. They could see coastlines and other ships miles away irrespective of the visibility. There was now an alternative to the lighthouse and the fog signal.

However, if anyone was foolish enough to believe that radar had eliminated the dangers of grounding or collision, they were sadly mistaken. On the evening of the 25th July 1956 the 29,000 ton Italian liner *Andrea Doria* collided with the 12,000 ton Swedish liner *Stockholm*, despite the fact that both ships were fully equipped with functioning radar equipment. Having the equipment was one thing, being able to use it properly was another. The sea is an environment where safety is achieved only by ceaseless care and vigilance and it was the lack of these qualities that led to the capsizing of the *Herald of Free Enterprise* on the 6th March 1987 when she left port with her bow doors open to the sea.

As the years have passed, ships have been fitted with ever more sophisticated navigational equipment and many have satellite-based systems that can identify their position to within a few metres. One ship that had all these devices and should, therefore, have been in no danger was the *Costa Concordia*, but the actions of Captain Francesco Schettino in January 2012 bear striking similarities to those of Captain Alexander Taylor of the *Margaret Smith* in 1879. Both men recklessly took their vessels too close to the shore with disastrous consequences, much worse in the case of the *Costa Concordia* as lives were lost.

Today, lighthouses are usually automated, light vessels have been replaced by large automatic buoys and for major vessels in particular, electronics has replaced the human senses. There are those who would argue that lighthouses are an anachronism and should be discontinued altogether. They would do well to remember that all their sophisticated navigational equipment is totally dependent upon electricity. If their electricity fails, then their navigational options are once again dependent upon their sense of sight. It is unsurprising, therefore, that it is the smaller craft that still make the most use of these aids to navigation.

Alexander George Findlay's contribution to safety at sea can never be measured, for if his work enabled a mariner to recognize danger and thus avoid it, then the event would never be recorded. His work lacked glamour, but it is certain that many thousands of sailors and the passengers who travelled on the ships they crewed have good reason to be grateful for his labours.

[90] The London International Lighthouse Conference, 1929 – HMSO, London, 1930.
[91] Report of the Proceedings of the International Technical Conference on Lighthouses and Other Aids to Navigation, Berlin, 1937 – HMSO, London 1939.

Bibliography

A Description and List of the Lighthouses of the World – Alexander George Findlay; Richard Holmes Laurie, London 1861.

A Description and List of the Lighthouses of the World, Fifth Edition 1865 - Alexander George Findlay; Richard Holmes Laurie, London, 1865.

A Description and List of the Lighthouses of the World, Seventh Edition 1867 – Alexander George Findlay; Richard Holmes Laurie, London, 1867.

A Description and List of the Lighthouses of the World, Nineteenth Edition – Alexander George Findlay; Richard Holmes Laurie, London, 1879.

A Description and List of the Lighthouses of the World, Twentieth Edition – Alexander George Findlay; Richard Holmes Laurie, London, 1880.

A Description and List of the Lighthouses of the World, Twenty-Seventh Edition – Alexander George Findlay; Richard Holmes Laurie, London, 1887.

A Description and List of the Lighthouses of the World, Thirtieth Edition – Alexander George Findlay; Richard Holmes Laurie, London, 1890.

A Description and List of the Lighthouses of the World, Thirty-Ninth Edition – Alexander George Findlay; Richard Holmes Laurie, London, 1899.

A Directory for the Navigation of the Indian Ocean with Descriptions of its Coasts Islands etc.- Alexander George Findlay; Richard Holmes Laurie, London, 1866.

A Few Notes on Modern Lighthouse Practice – Chance Brothers & Co. Ltd., Birmingham, 1910

A Rudimentary Treatise on the History, Construction and Illumination of Lighthouses – Alan Stevenson; John Weale, London, 1850.

A Short Bright Flash – Theresa Levitt; W. W. Norton, New York, 2013.

British Lighthouses, Their History and Romance – W. J. Hardy; Religious Tract Society, London, 1895.

Cornish Seafarers - A. K. Hamilton Jenkin; 1932.

Description and Plans of Lights for Lighthouses – Chance Brothers, Birmingham 1855.

Dictionary of National Biography.

Post Office Directory for Greenock 1879.

Report of the Proceedings of the International Technical Conference on Lighthouses and Other Aids to Navigation, Berlin, 1937 – HMSO, London 1939.

Lighthouses-The Race to Illuminate the World – Toby Chance and Peter Williams; New Holland, London, 2008.

Longitude – Dava Sobel; Harper Perennial, London 2011.

Moving Sunderland's Lighthouse 1841 – Ian Hills; FWD Publishing, 2019.

Neglectful or Worse – Cathryn Pearse in Troze, the Official Journal of the National Maritime Museum, Cornwall, 2008.

Optical Apparatus as used in Lighthouses – James T Chance; William Clowes and Sons, London 1867

Report from the Select Committee on Lighthouses - London 1845.

Royal Geographical Society Journal, Vol. XLV, 1875.

Sextant – David Barrie; William Collins, London, 2014.

The British Pharos – Alan Stevenson; W. Reid & Son, Leith, 1831.

The Fresnel Lens Makers – Thomas Tag; United States Lighthouse Society.

The Greatest Lie on Earth – Edward Hendrie; Great Mountain Publishing, 2016.

The Heyday of Sail – The Merchant Sailing Ship 1650-1830 – Conway Maritime Press, London, 1995.

The Lighthouse and the Battleship – Ian Hills; FWD Publishing, 2018.

The Lighthouse Work of Sir James Chance, Baronet – James Frederick Chance; Smith, Elder & Co., London 1902.

The Light-Houses of The British Isles, Hydrographical Office, Admiralty, London 1832.

The Lightkeeper – Gerald Butler; Liffey Press, Dublin, 2012.

The London International Lighthouse Conference, 1929 – HMSO, London, 1930.

The Practical Navigator – William Black; Greenock, 1839.

The Ship, The Life and Death of the Merchant Sailing Ship – Basil Greenhill; HMSO, London, 1980.

The Sinking of RMS Tayleur – Gill Hoffs; Pen and Sword, Barnsley, 2015.

The Story of the Belle Toute Lighthouse – Rob Wassell; RW Publications 2010.

The Unsinkable Titanic – Allen Gibson; History Press, Stroud, 2017.

Zetetic Astronomy – Samuel Birley Rowbotham; Pantianos Classics.

Printed in Great Britain
by Amazon